Milk and Milk Products Technology

Milk and Milk Products Technology

Dr. Pradeep Kumar Choda

RANDOM

RANDOM PUBLICATIONS

NEW DELHI (INDIA)

Milk and Milk Products Technology

ISBN 978-93-5111-793-3

Published in 2016 in India by

RANDOM PUBLICATIONS

4376-A/4B, Gali Murari Lal, Ansari Road
New Delhi-110 002
Phone : +9111-43580356, 011-23289044, 011-43142548
e-mail: sales@randompublications.com,
info@randompublications.com, randomexports@gmail.com

Reprinted 2019

Type Setting by : Friends Media, Delhi-110089
Digitally Printed at : Replika Press Pvt. Ltd.

Preface

Milk is a valuable nutritious food that has a short shelf-life and requires careful handling. Milk is highly perishable because it is an excellent medium for the growth of microorganisms - particularly bacterial pathogens - that can cause spoilage and diseases in consumers. Milk processing allows the preservation of milk for days, weeks or months and helps to reduce food-borne illness. The usable life of milk can be extended for several days through techniques such as cooling (which is the factor most likely to influence the quality of raw milk) or fermentation. Pasteurization is a heat treatment process that extends the usable life of milk and reduces the numbers of possible pathogenic microorganisms to levels at which they do not represent a significant health hazard. Milk can be processed further to convert it into high-value, concentrated and easily transportable dairy products with long shelf-lives, such as butter, cheese and ghee.

Fermented milk products, also known as cultured dairy foods, cultured dairy products, or cultured milk products, are dairy foods that have been fermented with lactic acid bacteria such as Lactobacillus, Lactococcus, and Leuconostoc. The fermentation process increases the shelf-life of the product, while enhancing the taste and improving the digestibility of milk. There is evidence that fermented milk products have been produced since around 10,000 BC. A range of different Lactobacilli strains has been grown in laboratories allowing for a wide range of cultured milk products with different tastes.

Milk and its by-products are what we call lactic (or dairy) products. The nutritional value of milk and its by-products mean that they are recommended for consumption at every stage of life (in infancy, youth, middle age and old age). Through milk, we can obtain a series of other food products such as butter, cheese, yoghurt and cream.

– *Author*

Contents

1

Milk and Milk Products

INTRODUCTION

Although cow's milk is the most popular in many countries, milk can be obtained from many different sources. For example, milk from goats and sheep makes a substantial contribution to the total milk production in countries of Eastern and Southern Europe, Malawi, and Barbados, whereas the water buffalo is a common source of milk in much of Asia.

Children and adolescents should be encouraged to consume milk products because this is the period of their lives in which they are building their peak bone mass and developing lifelong habits. This article highlights the benefits of milk and milk products, as well as common misconceptions.

WHAT ARE DAIRY PRODUCTS?

A production plant for processing dairy products is called a dairy or a dairy factory. Dairy products are generally defined as food products that are produced from milk. They are rich sources of energy. Raw milk for processing generally comes from cows, but occasionally from other mammals such as goats, sheep, and water buffalo. Water is the main constituent of milk (about 90%).

SOURCES OF DAIRY PRODUCTS

Milk of various types (including whole milk, skim milk, buttermilk), yoghurt, cheese (e.g. Swiss cheese, cheddar cheese, cottage cheese), and ice cream are dairy products. Of all milk products, milk, yoghurt and cheese are the best sources of calcium.

Nutritional value of dairy

Dairy products and alternatives such as calcium-fortified soy products are nutritious foods, and provide benefit when consumed as part of a nutritionally balanced diet which includes all of the 5 food groups:

- Breads and cereals;
- Vegetables and fruits;

- Dairy products/alternatives;
- Meat/chicken/fish/alternatives; and
- A small amount of fats and oils.

The functions of a food are served specifically through its nutritionally important components, includingproteins, carbohydrates, lipids, minerals, vitamins and water. Cow's milk is the preferred choice for most people. It provides 67 kilocalories and has a protein content of 3.2 grams per 100 millilitres. Milk proteins include casein (about 80%) and whey (about 20%). Whey has a higher nutritional value than casein.

Once fat and casein have been removed from milk, it consists mainly of whey, which contains the soluble milk salts, milk sugar and the remainder of the milk proteins. Whey proteins consist of a number of specialised proteins, the most important being beta lactoglobulin (50% of whey) and lactoglobulin.

Milk proteins have a high biological value but, unlike egg proteins, they lack sulphur-containing amino acids. The proteins in cow's milk have balanced amino acid profiles and good digestibility, making it the obvious choice when it comes to feeding the family. Casein in cow's milk combines with calcium to form caseinogen. A higher proportion of calcium and casein in cow's milk makes it more difficult to digest than human milk. Some people cannot drink fresh milk because they are lactose intolerant, but can consume sour milk because it contains less lactose.

The nutrient composition of milk makes it a time-tested liquid that is indispensable for the maintenance of optimal nutrition, especially for the young. The major nutrients in milk aside from protein include the following:

Fat

Milk and milk products also contain fat. Cow's milk contains fat that is in the form of glycerides. The fat in cow's milk is a poor source of essential fatty acids.

The fat content of milk varies:

- Whole milk contains 3.9 g fat per 100 ml,
- Semi-skimmed milk provides 1.7 g fat per 100 ml
- Skim milk provides 0.2 g fat per 100 ml
- 1% milk, a blend of skimmed and semi-skimmed milk, has recently become available. It contains 1 g fat or less per 100 ml.

Carbohydrates

The main carbohydrate in milk is a disaccharide called lactose. It is made up of two simple sugars –glucose and galactose – and is sweeter than sucrose (found in cane sugar). Lactose supports the absorption of calcium and phosphorus and the synthesis of some B complex vitamins in the small intestine. In the human body, an intestinal enzyme called lactase digests lactose. Some people are unable to produce enough lactase, inhibiting lactose digestion. This undigested lactose is then broken down in the large intestine by bacteria, causing the formation of gas, bloating, pain and diarrhoea. This condition is called lactose intolerance.

Lactose intolerance is the inability to digest significant amounts of lactose, the major sugar found in milk. Lactose intolerance poses a problem for some people, but can be managed. People who have trouble digesting lactose should avoid dairy products that are not compatible with their system. Many people with this condition can enjoy milk, ice cream and other such products if they eat them in small amounts or eat other food at the same time. Additionally, lactase liquid or tablets can help digest the lactose.

Lactase deficient individuals may tolerate milk with cereals or cooked as custard more than liquid milk. Some dairy foods contain less lactose than others, and may be better for people who suffer from lactose intolerance. These include:

- Fermented milk products, including some yoghurts, mature cheeses (like cheddar cheese) and butter, generally pose no problems.
- Since heating breaks down some of the lactose, heated milk products such as evaporated milk may be preferred to unheated foods.

Vitamins and minerals

Milk and milk products are an excellent source of vitamins and minerals, particularly calcium. Milk has significant amounts of vitamin A and B vitamins such as thiamine, riboflavin and nicotinic acid, but is a poor source of vitamin C and vitamin E. It also contains vitamin B12.

The important minerals in milk are calcium, phosphorus, sodium and potassium. It is, however, a poor source of iron. Calcium is a mineral that the body needs for numerous functions, including building and maintaining bones and teeth, blood clotting, the transmission of nerve impulses, and regulation of theheart's rhythm.

WOMAN MILKING A GOAT

Milk is a perishable commodity and spoils very easily. Its low acidity and high nutrient content make it the perfect breeding ground for bacteria, including those which cause food poisoning (pathogens).

Bacteria from the animal, utensils, hands, and insects may contaminate the milk, and their destruction is the main reason for processing. This preservation of the milk can be achieved by fermentation, heating, cooling, removal of water, and by concentration or separation of components, to produce foods such as butter or cheese.

The degree to which milk consumption and processing occurs will differ from region to region. It is dependent upon a whole host of factors, including geographic and climatic conditions, availability and cost of milk, food taboos, and religious restrictions. Where processing does exist, many traditional techniques can be found for producing indigenous milk products. These are more stable than raw milk and provide a means of preservation as well as adding variety to the diet. In addition, the introduction of western-style dairy products and the subsequent setting up of small-scale dairies has provided more choice of dairy products to the consumer.

MILK AND ITS PRODUCTS

Milk is a complete food for the young animal because it contains the five fundamental types of food material -water, mineral matter, fat, carbohydrate, and protein.

The analysis of average milk is about as follows:

	Per cent.
Water......	.87
Mineral........	.01
Fat	.04
Casein........	.03
Sugar.	.05
	1.00

Since, the fat is the most valuable portion commercially, dairymen study to feed their cows in such a way as to increase it, and in some instances milk has been produced containing 6 per cent of fat.

Though mainly water, milk is a valuable nutritious food and should be used freely by itself and in combination with other food materials, insoups, sauces, and puddings.

When we remember what the department of agriculture has proved for us, that a quart ofmilkis quite as nourishing as a quart of oysters for which we pay six or eight times as much, we can see that it is desirable to use it more freely than is generally done. Especially during the summer months we do well to substitute milk andcheeseformeats. There are average families which do not use over a pint of milk a day; there are others who find it necessary to take a gallon, and the meat bill in the latter cases becomes proportionately small. A pint of milk a day is not an excessive allowance for each member of a family, though many households consume much less.

To study the composition of milk put a quart of fresh milk in aglass jarand leave it twenty-four hours or longer until it is thick and sour. Whatpercentage ofthe whole is the cream? Remove the layer of cream on top to another jar, screw on the top, and shake until the fat separates from the watery portion of the milk. Collect thebutteron a spoon, wash out the milk by pressing and folding with a knife.

Weigh or estimate carefully the value of the butter obtained. What proportion of the original bulk of milk does it represent? Persons fond of unsalted butter may thus prepare it for themselves.

WHY IS SALT ADDED TO BUTTER?

The remainder of the milk, now a thick mass of curd, may be pressed out with a spoon or cut with a knife to show the greenish water known as whey. What nutritive substances are there in this?

Turn the thick milk into a two-quart pan and fill with hot water, in twenty minutes drain the water off through a strainer, that no curd need be lost, and pour on more hot water. Do this several times until the curd loses its sour taste and has contracted, but do not allow it to become too hard. If boiling water is used the curd will become unpalatable and indigestible.

Compositi of Milk

Buttons have been made of sour milk treated by heat and pressure.

Press as much water as possible from the curd and compare the quantity with the original amount of milk. Remember that this still contains much water. Now combine the curd with butter or thick cream, salt it and shape in small balls or pack in cups. Thus we learn something of the value of milk and have made a sour milk cheese more palatable than when the whole mass of curdled milk is heated on the stove or strained in a cloth.

With prepared rennet in liquid or tabletformthe curd and whey of sweet milk may be separated. The milk should be warmed slightly before the dissolved rennet is added, then chilled in the dishes from which it is to be served. This is known asjunketor rennet custard.

Absolute cleanliness is essential for every utensil to come in contact with milk. The souring of the milk is due to the action of bacteria which come to it from contact with utensils and the air. Its fluid form and nutritive material afford a medium peculiarly favorable to the development of germs of disease, as well as to the growth of useful bacteria which aid in butter and cheese making. The growth of such micro-organisms is hastened by moderate heat, but most of them are killed by raising the milk to the boiling point.

Sterilization requires atemperatureof two hundred and twelve degrees F, continued for about twenty minutes; this process usually changes the flavour of the milk so that it is disagreeable to many palates. The high temperature also causes the fat globules to separate instead of being retained inthe formof cream. Pasteurization takes its name from the noted French scientist, and consists in raising the milk to a temperature of about one hundred and fifty-five degrees F. By this means the flavour of the milk is unchanged.

The cook finds it safe to scald the milk for soups,bread, or puddings, to prevent its souring during the process, beforecookingit with the other ingredients. There is a gain in the time of cooking when the milk-is heated while the other materials are being prepared. A bit of bicarbonate of soda dissolved in milk before it is heated often will neutralise any incipient acidity and make it usable for puddings or soups. The "cream" oftomato soupis liable to curdle unless the acid of the tomato is neutralised by soda or the milk thickened with flour before the two parts are combined. It is safer with all "cream" soups to keep thestockand thickened milk apart until just before using.

Lemon or other acid fruit juices are sometimes mixed with milk for sherbet without curdling if, before the juice is added, the milk is thoroughly chilled in the freezer can.

- To Prevent Souring
- Neutralizin Acidity
- Mixing with Acid.

Salt sometimes curdles milk, especially when it is added to hot milk. Since, the solid portions of milk readily adhere to the bottom of the saucepan placed in direct contact with heat, and the resulting burned flavour rapidly penetrates the whole of the milk, adouble boileror its equivalent, onedish setin another of boiling water, is the best way to heat milk. Milk is an important ingredient in preparing cocoa andchocolate, and such beverages rank with soup in nutritive value. Hot milk sipped slowly is a simple remedy for exhaustion and sleeplessness. Hot milk should be served with coffee when cream is not available. The milk soups are valuable foods and have as their foundation thewhite saucedescribed further on. Most of our puddings require milk, especially the cereala nd custard varieties.

Because there are solids in the milk more time must be allowed for thegrainsof rice or corn meal to absorb the moisture than when cooked in water. The protein portions of the milk have somewhat the same effect as the egg used to coat the croquette or oyster before frying. If the particles of grain are thus varnished over they cannot absorb moisture as rapidly as from clear water. Hence, it is often advisable to cook the grains in water first and finish the process in the milk. In making blanc mange from Irish moss, if the moss is first cooked in a small quantity of water and the thick paste strained before it is added to the milk, there is no loss of milk. When the moss is cooked directly in the milk there is some loss of milk when the moss is strained out. The baked Indian meal pudding and the creamy rice pudding require long, gentle baking. There is a continual evaporation of moisture from the surface of the pudding pan, and really a condensing of the milk. In proportion as the pudding dish is refilled with milk, the pudding increases in nutritive value. Milk is commonly used for mixing dough of many types and this adds to the nutritive value of bread and cakes.

Bread made of milk or part milk will have a browner, tenderer crust than bread made wholly with water. There seems to be good ground, however, for the prevalent idea that bread or cake made with milk does not keep so well as that made with water. A certain cheesy flavour develops where milk is a principal ingredient. Sour milk is often used for mixing griddle cakes andquick doughs, because the acid it contains will be neutralised by the soda added, and thus produce the effervescence which makes the dough light. The souring process seems to have so affected the protein substances in the milk that such a dough is tenderer than one made with sweet milk andbaking powder. The use of sour milk will be further treated in the section on doughs. For doughs, soups, and puddings, in which additional fat is introduced, skimmed milk may be used as well as full milk. The use of cream in well-to-do families is increasing. Whipped cream is demanded as a garnish or sauce for many desserts quite complete in themselves. The process of beating or "whipping" cream gives it an attractive appearance, and by expanding its particles probably makes it more digestible.

NUTRITIONAL SIGNIFICANCE OF MILK

Milk is often regarded as being nature's most complete food. It earns this reputation by providing many of the nutrients which are essential for the growth of the human body. Being an excellent source of protein and having an abundance of vitamins and minerals, particularly calcium, milk can make a positive contribution to the health of a nation. The realisation of its nutritional attributes is clearly illustrated by the implementation of numerous 'school milk programmes' worldwide. Fermented-milk products such as yoghurt and soured milk contain bacteria from the *Lactobacilli* group. These bacteria occur naturally in the digestive tract and have a cleansing and healing effect. Therefore the introduction of fermented products into the diet can help prevent certain yeasts and bacteria which may cause illness.

Many people suffer from a condition known as 'lactose intolerance'. This means that they are unable to digest the milk fat (lactose). Such people can, however, tolerate milk if it is fermented to produce foods such as yoghurt. During fermentation, lactic acid producing bacteria break down lactose, and in doing so eliminate the cause of irritation.

NUTRIENT COMPOSITION

Although milk is a liquid and most often considered a drink, it contains between 12 and 13 percent total solids and perhaps should be regarded as a food. In contrast, many "solid" foods, such as tomatoes, carrots, and lettuce, contain as little as 6 percent solids. Many factors influence the composition of milk, including breed, genetic constitution of the individual cow, age of the cow, stage of lactation, interval between milkings, and certain disease conditions. Since the last milk drawn at each milking is richest in fat, the completeness of milking also influences a sample. In general, the type offeed only slightly affects the composition of milk, but feed of poor quality or insufficient quantity causes both a low yield and a low percentage of total solids. Current feeding programs utilize computer technology to achieve the greatest efficiency from each animal.

The composition of milk varies among mammals, primarily to meet growth rates of the individual species. The proteins contained within the mother's milk are the major components contributing to the growth rate of the young animals. Human milk is relatively low in both proteins and minerals compared with that of cows and goats.

Goat milk has about the same nutrient composition as cow milk, but it differs in several characteristics. Goat milk is completely white in colour because all the beta-carotene (ingested from feed) is converted to vitamin A. The fat globules are smaller and therefore remain suspended, so the cream does not rise and mechanicalhomogenization is unnecessary. Goat milk curd forms into small, light flakes and is more easily digested, much like the curd formed from human milk. It is often prescribed for persons who are allergic to the proteins in cow milk and for some patients afflicted with stomach ulcers.

Sheep milk is rich in nutrients, having 18 percent total solids (5.8 percent protein and 6.5 percent fat). Reindeer milk has the highest level of nutrients, with 36.7 percent total solids (10.3 percent protein and 22 percent fat). These high-fat, high-protein milks are excellent ingredients for cheese and other manufactured dairy products. The major components of milk are water, fat, protein, carbohydrate (lactose), and minerals (ash). However, there are numerous other highly important micronutrients such as vitamins, essential amino acids, and trace minerals. Indeed, more than 250 chemical compounds have been identified in milk.

Table. Nutrient composition of dairy products (per 100 g)

dairy product	energy (kcal)	water (g)	protein (g)	fat (g)	carbohy-drate (g)	cholesterol (mg)	vitamin A (IU)	riboflavin (mg)	calcium (mg)
fresh milk									
whole	61	88	3.29	3.34	4.66	14	126	0.162	119
low-fat*	50	89	3.33	1.92	4.80	8	205	0.165	122
skim*	35	91	3.41	0.18	4.85	2	204	0.140	123
evaporated milk	134	74	6.81	7.56	10.04	29	243	0.316	261
evaporated skim milk*	78	79	7.55	0.20	11.35	4	392	0.309	290
sweetened condensed milk	321	27	7.91	8.70	54.40	34	328	0.416	284
nonfat dry milk*	358	4	35.10	0.72	52.19	18	2,370	1.744	1,231
butter	717	16	0.85	81.11	0.06	219	3,058	0.034	24
ice cream (vanilla)	201	61	3.50	11.00	23.60	44	409	0.240	128
ice milk (vanilla)	139	68	3.80	4.30	22.70	14	165	0.265	139
sherbet (orange)	138	66	1.10	2.00	30.40	5	76	0.068	54
frozen yogurt, nonfat	128	69	3.94	0.18	28.16	2	7	0.265	134
buttermilk	40	90	3.31	0.88	4.79	4	33	0.154	116
sour cream	214	71	3.16	20.96	4.27	44	790	0.149	116
yogurt, plain, low-fat	63	85	5.25	1.55	7.04	6	66	0.214	183
yogurt, fruit, low-fat	102	74	4.37	1.08	19.05	4	46	0.178	152
Cheese									
blue	353	42	21.40	28.74	2.34	75	721	0.382	528
Brie	334	48	20.75	27.68	0.45	100	667	0.520	184
Cheddar	403	37	24.90	33.14	1.28	105	1,059	0.375	721
cottage	103	79	12.49	4.51	2.68	15	163	0.163	60
cream	349	54	7.55	34.87	2.66	110	1,427	0.197	80
mozzarella**	280	49	27.47	17.12	3.14	54	628	0.343	731
Parmesan, grated	456	18	41.56	30.02	3.74	79	701	0.386	1,376
Emmentaler (Swiss)	376	37	28.43	27.54	3.38	92	845	0.365	961

*Fortified with vitamin A.
**Low moisture, part skim.

The quality of milk

The type of animal, its quality, and its diet can lead to differences in the colour, flavour, and composition of milk. Infections in the animal which cause illness may be passed directly to the consumer through milk. It is therefore extremely important that quality-control tests are carried out to ensure that the bacterial activity in raw milk is of an acceptable level, and that no harmful bacteria remain in the processed products.

Table. Average composition (%) of milks of various mammals

Species	*Water*	*Fat*	*Protein*	*Lactose*	*Ash*
Human	87.43	3.75	1.63	6.98	0.21
Cow	87.2	3.7	3.5	4.9	0.7
Goat	87.00	4.25	3.52	4.27	0.86
Sheep	80.71	7.9	5.23	4.81	0.9
Indian buffalo	82.76	7.38	3.6	5.48	0.78
Camel	87.61	5.38	2.98	3.26	0.7
Horse	89.04	1.59	2.69	6.14	0.51
Llama	86.55	3.15	3.9	5.6	0.8

STANDARD TESTING PROCEDURES

Milk fat

The price paid for milk is usually dependent upon the milk-fat content, and this may be determined either at the collection stage or at the dairy using a piece of equipment known as a butyrometer. Additionally the specific gravity can be measured using a hydrometer. This can also be used as an aid to detect adulteration.

Bacterial activity

Routinely it is necessary to check the microbiological quality of raw milk using either methylene blue or resazurin dyes. These tests indicate the activity of bacteria in the milk sample and the results determine whether the milk is accepted or rejected. Both tests work on the principle of the time taken to change the colour of the dye. The length of time taken is proportional to the number of micro-organisms present (the shorter the time taken, the higher the bacterial activity). It is preferable to use the resazurin test as this is less time-consuming. For these tests, basic laboratory equipment will be needed such as test-tubes, a water bath, accurate measuring equipment, and a supply of dyes. After collection the milk should ideally be stored at a temperature of 4°C or below. This is necessary to slow the growth of any contaminating bacteria.

Phosphatase test

For pasteurized milk, it is possible to ensure that pasteurization has been adequately achieved by testing for the presence of the enzyme phosphatase.

The destruction of phosphatase is regarded as a reliable test to show that the milk has been sufficiently heat-processed, because this enzyme (present in raw milk) is destroyed by pasteurization conditions. It is stressed that pasteurization is an effective safeguard against spoilage and food poisoning only if the milk is not re-contaminated after pasteurization.

PROCESSING OF LIQUID MILK

Milk can be kept for longer periods of time if it is heated to destroy the bacteria or cooled to slow their growth. Pasteurization and sterilization are the two most commonly-used heat treatments. Technically, it is possible for both to be carried out on a small scale, but they are most usually performed on a larger industrial scale due to the need for qualified, experienced staff and accurate and strictly controlled hygienic processing conditions.

Table. Production stages for pasteurized and sterilized milk

Product	*Store raw milk at 4°C*	*Test for bacterial activity using resazurin/methylene blue*	*Filter*	*Homogenize*	*Pasteurize*	*Fill into sterilized bottles*	*Sterilize*	*Store*
Pasteurized milk	*	*	*		*	*		*
Sterilized milk	*	*	*	*		*	*	*

Table. Equipment required

Processing stage	*Equipment*
Store at 4°C	Refrigerated storage
	Thermometer
Test for fat content	Butyrometer
Test specific gravity	Hydrometer
Test bacterial activity	Supply of dyes
	Thermometer Basic laboratory equipment is required for most of the tests
Filter	Filter cloth
	Filter press
Homogenization	Homogenizer
Fill into bottles	Liquid-filling machine
	Capping machine
Pasteurization	Boiling pan
	or pasteurizer
	Heat source
	Thermometer
Sterilization	Pressure cooker
	Retort
	Heat source
	Thermometer
Cool	Bottle-cooling system

Homogenization

Homogenization breaks up the oil droplets in milk and prevents the cream from separating out and forming a layer. This is of particular importance for sterilized milk which has a long shelf-life and when the formation of a cream layer is not desired. Additional changes include increased viscosity and a richer taste. Homogenizers are more usually designed for industrial-scale production, but it is possible to purchase smaller versions.

Filling

The most common packaging material for both pasteurized and sterilized milk is glass bottles sealed with either foil or metal caps, although plastic bottles, plastic bags, and cardboard cartons are all used when bottles are not available or too expensive.

Pasteurization

Pasteurization is a relatively mild heat treatment, (usually performed below 100°C) which is used to extend the shelf-life of milk for several days. It preserves the milk by the inactivation of enzymes and destruction of heat-sensitive micro-organisms, but causes minimal changes to the nutritive value or sensory characteristics of a food. Some heat-resistant bacteria survive to spoil the milk after a few days, but these bacteria do not cause food poisoning.

The time and temperature combination needed to destroy 'target' microorganisms will vary according to a number of complex inter-related factors. For milk, the heating time and temperature is either 63°C for 30 minutes or alternatively 72°C for 15 seconds. Only the former combination is possible on a small scale and for this the simplest equipment required is an open boiling pan. Better control is achieved using a steam jacketed pan, and this can be fitted with a stirrer to improve the efficiency of heating. Both of these are batch processes which are suited to small-scale operation. A higher production rate may be possible using a tubular-coil pasteurizer. This equipment has been tested and has been successful for some fruit products but it is presently still at a developmental stage.

Sterilization

Sterilization is a more severe heat treatment designed to destroy all contaminating bacteria. The milk is sterilized at a temperature of 121°C maintained for 15-20 minutes. This can be achieved using a retort or pressure cooker. Unlike pasteurization, this process causes substantial changes to the nutritional and sensory quality of the milk. In some countries, flavoured milk has become a very popular product. However, sterilization is not recommended for small-scale production for the following reasons:

- The cost of a retort and ancillary equipment is high for the small-scale processor.
- It is essential that the correct heating conditions are carefully established and maintained for every batch of milk that is processed.

If the milk is overheated, the quality is reduced, and it may have a rather burnt taste and aroma.

- If the milk is not heated sufficiently, there is a risk that micro-organisms will survive and grow inside the bottle. In low-acid foods such as milk, many types of bacteria including *Clostridium botulinum* can grow and cause severe food poisoning.
- Due to the potential dangers from food poisoning, the skills of a qualified food technologist/microbiologist are required in order to routinely examine samples of sterilized milk that have been subjected to accelerated storage conditions. This requires a supply of microbiological media and equipment.

The process of sterilization requires a considerable capital investment, the need for trained and experienced staff, regular maintenance of sophisticated equipment, and a comparatively high operating expenditure.

Cooling

Pasteurization does not destroy all of the micro-organisms, therefore the milk has to be cooled rapidly to prevent the growth of surviving bacteria. Cooling can be achieved on a small scale by using a bottle-cooling system.

Storage

Pasteurized milk has a shelf-life of 2-3 days if kept at 4°C. Maintaining this low temperature causes a substantial increase to the cost of transportation and distribution and is therefore a major disadvantage to the development of a small-scale pasteurized milk business. If packaged in sealed bottles and stored at room temperature, sterilized milk should have a shelf-life in excess of six months.

PHYSICAL AND BIOCHEMICAL PROPERTIES

Milk contains many natural enzymes, and other enzymes are produced in milk as a result of bacterial growth. Enzymes are biological catalysts capable of producing chemical changes in organic substances. Enzyme action in milk systems is extremely important for its effect on the flavour and body of different milk products. Lipases(fat-splitting enzymes), oxidases, proteases (protein-splitting enzymes), and amylases(starch-splitting enzymes) are among the more important enzymes that occur naturally in milk. These classes of enzymes are also produced in milk by microbiological action. In addition, the proteolytic enzyme (i.e., protease) rennin, produced in calves' stomachs to coagulate milk protein and aid in nutrient absorption, is used to coagulate milk for manufacturing cheese.

The coagulation of milk is an irreversible change of its protein from a soluble or dispersed state to an agglomerated or precipitated condition. Its appearance may be associated with spoilage, but coagulation is a necessary step in many processing procedures. Milk may be coagulated by rennin or other enzymes, usually in conjunction with heat. Left unrefrigerated, milk may naturally sour or coagulate by the action of lactic acid, which is produced by lactose-fermenting bacteria. This principle is utilized in the manufacture of cottage cheese. When milk is pasteurized and continuously refrigerated for two or three weeks, it may eventually coagulate or spoil owing to the action of psychrophilic or proteolytic organisms that are normally present or result from postpasteurization contamination.

Milk fat is present in milk as an emulsion in a water phase. Finely dispersed fat globules in this emulsion are stabilized by a milk protein membrane, which permits the fat to clump and rise. The rising action is called creaming and is expected in all unhomogenized milk. In the United States, when paper cartons supplanted glass bottles, consumers stopped the practice of skimming cream from the top. Processors then introduced homogenization, a method of preventing gravity separation by forcing milk through very small openings under pressure, thus reducing fat globules to one-tenth their original size. Homogenization is practiced in many dairy processes in order to improve the physical properties of products.

Milk and other dairy products are very susceptible to developing off-flavours. Some flavours, given such names as "feed," "barny," or "unclean," are absorbed from the food ingested by the cow and from the odours in its surroundings. Others develop through microbial action due to growth of bacteria in large numbers. Chemical changes can also take place through enzyme action, contact with metals (such as copper), or exposure to sunlight or strong fluorescent light. Quality-control directors are constantly striving to avoid off-flavours in milk and other dairy foods.

Fresh fluid milk

Fresh fluid milk requires the highest-quality raw milk and is generally designated as Grade "A." This grade requires a higher level of sanitation and inspection on the farm than is necessary for "manufacturing grade" milk.

Quality concerns

Raw milk is a potentially dangerous food that must be processed and protected to assure its safety for humans. While most bovine diseases, such as brucellosis and tuberculosis, have been eliminated, many potential human pathogens inhabit the dairy farm environment. Therefore, it is essential that all milk be either pasteurized or (in the case of cheese) held for at least 60 days if made from raw milk. While milk from healthy cows is often totally bacteria-free, that condition quickly changes when milk is exposed to the farm environment.

Milk received at the processing plant is tested before being unloaded from either farm-based tank trucks or over-the-road tankers. The milk is checked for odour, appearance, proper temperature, acidity, bacteria, and the presence of drug residues.

These tests take no longer than 10 to 15 minutes. If the tank load passes these tests, the milk is pumped into the plant's refrigerated storage tanks. The milk is then stored for the shortest possible time.

PHYSICAL STATUS OF MILK

About 87% of milk is water, in which the other constituents are distributed in various forms. We distinguish among several kinds of distribution according to the type and size of particle present in the liquid.

Kind of solution	Particle diameter (nm)
Ionic solution	0.01–1
Molecular solution	0.1–1
Colloid (fine dispersion)	1–100
Coarse dispersion (suspension or emulsion)	50–100

In milk we find examples of emulsions, colloids, molecular and ionic solutions.

IONIC SOLUTIONS

An ionic solution is obtained when the forces that hold the ions together in a solid salt are overcome. The dissolved salt breaks up into ions which float freely in the solvent.

Thus when common salt—sodium chloride—is dissolved in water it becomes an ionic solution of free sodium and chloride ions. Ionic solutions are largely of inorganic compounds.

MOLECULAR SOLUTIONS

In a molecular solution the molecules are only partly, if at all, dissociated into ions. The degree of dissociation represents an equilibrium which is influenced by other substances in the solution and by the pH (or hydrogen ion concentration) of the solution. Molecular solutions are usually of organic compounds

Colloids

In a colloid, one substance is dispersed in another in a finer state than an emulsion but the particle size is larger than that in a true solution. Colloidal systems are classified according to the physical state of the two phases. In a colloid, solid particles consisting of groups of molecules float freely. The particles in a colloid are much smaller than those in a suspension and a colloid is much more stable.

Emulsions

An emulsion consists of one immiscible liquid dispersed in another in the form of droplets—the disperse phase. The other phase is referred to as the continuous phase. The systems have minimal stability and require the presence of a surface-active or emulsifying agent for stability. In foods, emulsions usually contain oil and water. If water is the continuous phase and oil the disperse phase, it is an oil-in-water (o/w) emulsion, e.g. milk or cream. In the reverse case the emulsion is a water-in-oil (w/o) type, e.g. butter. In summary, an emulsion consists of three elements, the continuous phase, the disperse phase and the emulsifying agent.

Dispersions

A dispersion is obtained when particles of a substance are dispersed in a liquid. A suspension consists of solid particles dispersed in a liquid, and the force of gravity can cause them to sink to the bottom or float to the top. For example, fine sand, dispersed in water, soon settles out.

pH and acidity

An acid is a substance which dissociates to produce hydrogen ions in solution. A base (alkaline) is a substance which produces hydroxyl ions in solution. It can equally be stated that an acid is a substance which donates a proton and a base is a substance which accepts a proton.

The symbol pH is used to denote acidity; it is inversely related to hydrogen ion concentration.

Neutrality is pH 7

Acidity is less than pH 7

Alkalinity is more than pH 7

Fresh milk has a pH of 6.7 and is therefore slightly acidic.

When an acid is mixed with a base, neutralisation takes place; similarly a base will be neutralised by an acid.

BUFFER SOLUTIONS

Buffers are defined as materials that resist a change in pH on addition of acid or alkali. Characteristically they consist of a weak acid or a weak base and its salt. Milk contains a large number of these substances and consequently behaves as a buffer solution.

Fresh cows milk has a pH of between 6.7 and 6.5. Values higher than 6.7 denote mastitic milk and values below pH 6.5 denote the presence of colostrum or bacterial deterioration. Because milk is a buffer solution, considerable acid development may occur before the pH changes. A pH lower than 6.5 therefore indicates that considerable acid development has taken place. This is normally due to bacterial activity. Litmus test papers, which indicate pH, are used to test milk activity; pH measurements are often used as acceptance tests for milk. Measuring milk acidity is an important test used to determine milk quality. Acidity measurements are also used to monitor processes such as cheese-making and yoghurt-making.

The titratable acidity of fresh milk is expressed in terms of percentage lactic acid, because lactic acid is the principal acid produced by fermentation after milk is drawn from the udder and fresh milk contains only traces of lactic acid.

However, due to the buffering capacity of the proteins and milk salts, fresh milk normally exhibits an initial acidity of 0.14 to 0.16% when titrated using sodium hydroxide to a phenolphthalein end-point.

MILK CONSTITUENTS

The quantities of the main milk constituents can vary considerably depending on the individual animal, its breed, stage of lactation, age and health status. Herd management practices and environmental conditions also influence milk composition. The average composition of cows milk.

Table. *Composition of cows milk*

Main constituent	Range (%)	Mean (%)
Water	85.5 – 89.5	87.0
Total solids	10.5 – 14.5	13.0
Fat	2.5 – 6.0	4.0
Proteins	2.9 – 5.0	3.4
Lactose	3.6 – 5.5	4.8
Minerals	0.6 – 0.9	0.8

Water is the main constituent of milk and much milk processing is designed to remove water from milk or reduce the moisture content of the product.

MILK FAT

If milk is left to stand, a layer of cream forms on the surface. The cream differs considerably in appearance from the lower layer of skim milk.

Under the microscope cream can be seen to consist of a large number of spheres of varying sizes floating in the milk. Each sphere is surrounded by a thin skin—the fat globule membrane—which acts as the emulsifying agent for the fat suspended in milk.The membrane protects the fat from enzymes and prevents the globules coalescing into butter grains. The fat is present as an oil-in-water emulsion: this emulsion can be broken by mechanical action such as shaking.

Fats are partly solid at room temperature. The term oil is reserved for fats that are completely liquid at room temperature. Fats and oils are soluble in non-polar solvents, e.g. ether.

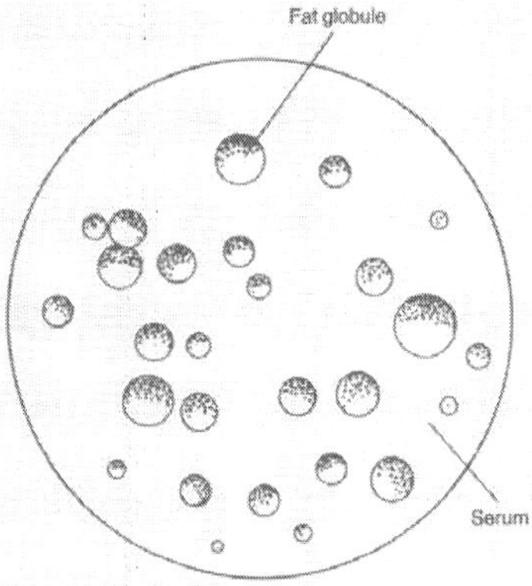

Fig.*Fat globules in milk.*

About 98% of milk fat is a mixture of triacyl glycerides. There are also neutral lipids, fat-soluble vitamins and pigments (e.g. carotene, which gives butter its yellow colour), sterols and waxes. Fats supply the body with a concentrated source of energy: oxidation of fat in the body yields 9 calories/g. Milk fat acts as a solvent for the fat-soluble vitamins A, D, E and K and also supplies essential fatty acids (linoleic, linolenic and arachidonic).

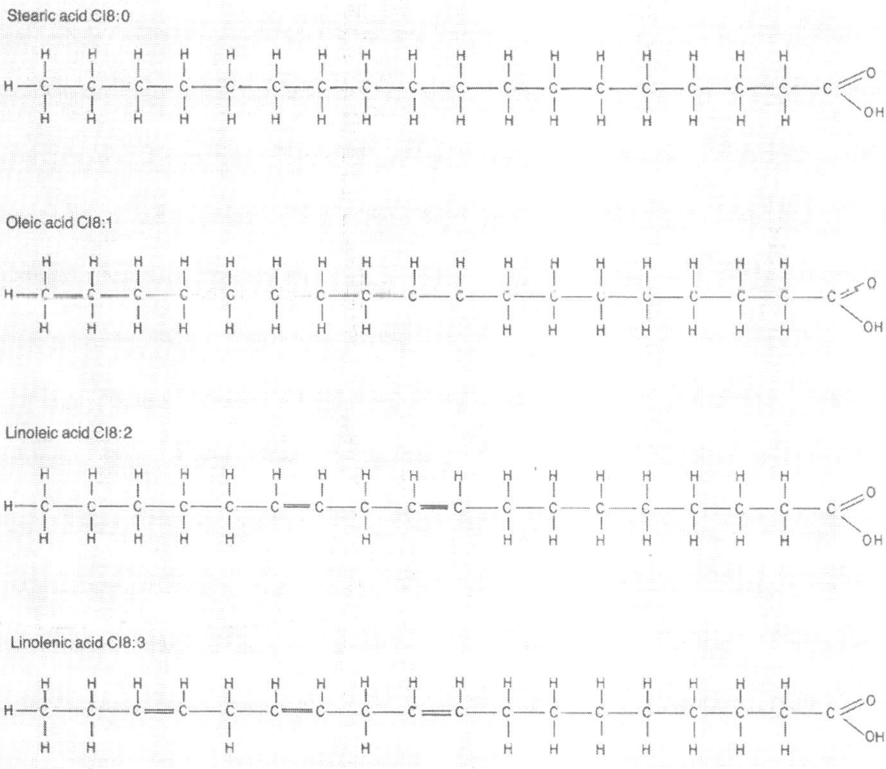

Fig. *Structural formulae of four 18-carbon fatty acids varying in degree of saturation.*

A fatty-acid molecule comprises a hydrocarbon chain and a carboxyl group (-COOH). In saturated fatty acids the carbon atoms are linked in a chain by single bonds. In unsaturated fatty acids there is one double bond and in poly-unsaturated fatty acids there is more than one double bond. Examples of each type of fatty acid.

Fatty acids vary in chain length from 4 carbon atoms, as in butyric acid (found only in butterfat), to 20 carbon atoms, as in arachidonic acid. Nearly all the fatty acids in milk contain an even number of carbon atoms.

Fatty acids can also vary in degree of unsaturation, e.g. C18:0 stearic (saturated), C18:1 oleic (one double bond), C18:2 linoleic (two double bonds), C18:3 linolenic (three double bonds).

The most important fatty acids found in milk triglycerides. Fatty acids are esterified with glycerol as follows:

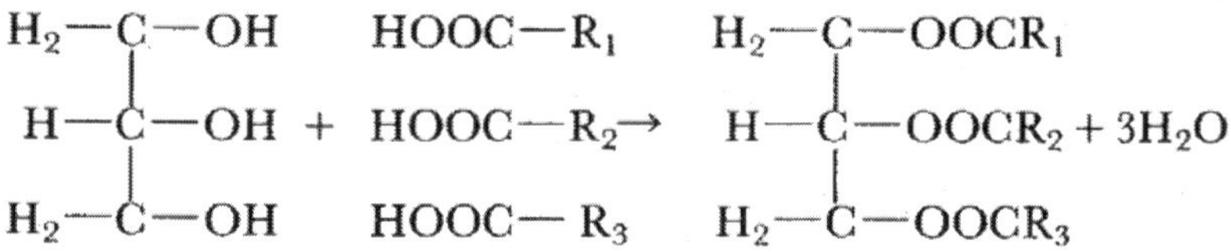

Glycerol + fatty acids → triglyceride (fat) + water

Table. ***Principal fatty acids found in milk triglycerides.***

	Molecular formula	Chain length	Melting point
Butyric	$CH_3(CH_2)_2COOH$	C_4	–8°C
Caproic	$CH_3(CH_2)_4COOH$	C_6	–2°C
Caprylic	$CH_2(CH_2)_6COOH$	C_8	16°C
Capric	$CH_3(CH_2)_8COOH$	C_{10}	31.5°C
Lauric	$CH_3(CH_2)_{10}COOH$	C_{12}	44°C
Myristic	$CH_3(CH_2)_{12}COOH$	C_{14}	58°C
Palmitic	$CH_3(CH_2)_{14}COOH$	C_{16}	64°C
Stearic	$CH_3(CH_2)_{16}COOH$	C_{18}	70°C
Arichidonic	$CH_3(CH_2)_{18}COOH$	C_{20}	
Oleic	$CH_3(CH_2)_7CH{=}CH(CH_2)_7COOH$	$C_{18:1}$	13°C
Linoleic	$CH_3(CH_2)_4(CH{=}CH.CH_2)_2(CH_2)_6COOH$	$C_{18:2}$	–5°C
Linolenic	$CH_3.CH_2(CH{=}CH.CH_2)_3(CH_2)_6COOH$	$C_{18:3}$	

The melting point and hardness of the fatty acid is affected by:

- The length of the carbon chain, and
- The degree of unsaturation.

As chain length increases, melting point increases. As the degree of unsaturation increases, the melting point decreases.

Fats composed of short-chain, unsaturated fatty acids have low melting points and are liquid at room temperature, i.e. oils. Fats high in long-chain saturated fatty acids have high melting points and are solid at room temperature. Butterfat is a mixture of fatty acids with different melting points, and therefore does not have a distinct melting point. Since, butterfat melts gradually over the temperature range of 0–40°C, some of the fat is liquid and some solid at temperatures between 16 and 25°C. The ratio of solid to liquid fat at the time of churning influences the rate of churning and the yield and quality of butter.

Fats readily absorb flavours. For example, butter made in a smoked gourd has a smokey flavour.

Fats in foods are subject to two types of deterioration that affect the flavour of food products.

1. *Hydrolytic rancidity:* In hydrolytic rancidity, fatty acids are broken off from the glycerol molecule by lipase enzymes produced by milk bacteria. The resulting free fatty acids are volatile and contribute significantly to the flavour of the product.
2. *Oxidative rancidity:* Oxidative rancidity occurs when fatty acids are oxidised. In milk products it causes tallowy flavours. Oxidative rancidity of dry butterfat causes off-flavours in recombined milk.

MILK PROTEINS

Proteins are an extremely important class of naturally occurring compounds that are essential to all life processes. They perform a variety of functions in living organisms ranging from providing structure to reproduction. Milk proteins represent one of the greatest contributions of milk to human nutrition. Proteins are polymers of amino acids. Only 20 different amino acids occur, regularly in proteins. They have the general structure:

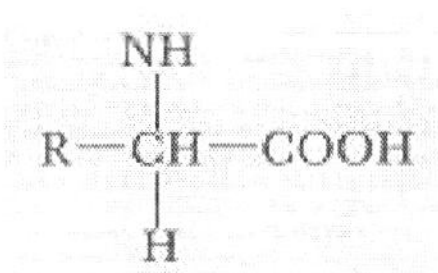

R represents the organic radical. Each amino acid has a different radical and this affects the properties of the acid. The content and sequence of amino acids in a protein therefore affect its properties. Some proteins contain substances other than amino acids, e.g. lipoproteins contain fat and protein. Such proteins are called conjugated proteins:

Phosphoproteins: Phosphate is linked chemically to these proteins—examples include casein in milk and phosphoproteins in egg yolk.

Lipoproteins: These combinations of lipid and protein are excellent emulsifying agents. Lipoproteins are found in milk and egg yolk.

Chromoproteins: These are proteins with a coloured prosthetic group and include haemoglobin and myoglobin.

CASEIN

Casein was first separated from milk in 1830, by adding acid to milk, thus establishing its existence as a distinct protein. In 1895 the whey proteins were separated into globulin and albumin fractions.

It was subsequently shown that casein is made up of a number of fractions and is therefore heterogeneous. The whey proteins are also made up of a number of distinct proteins.

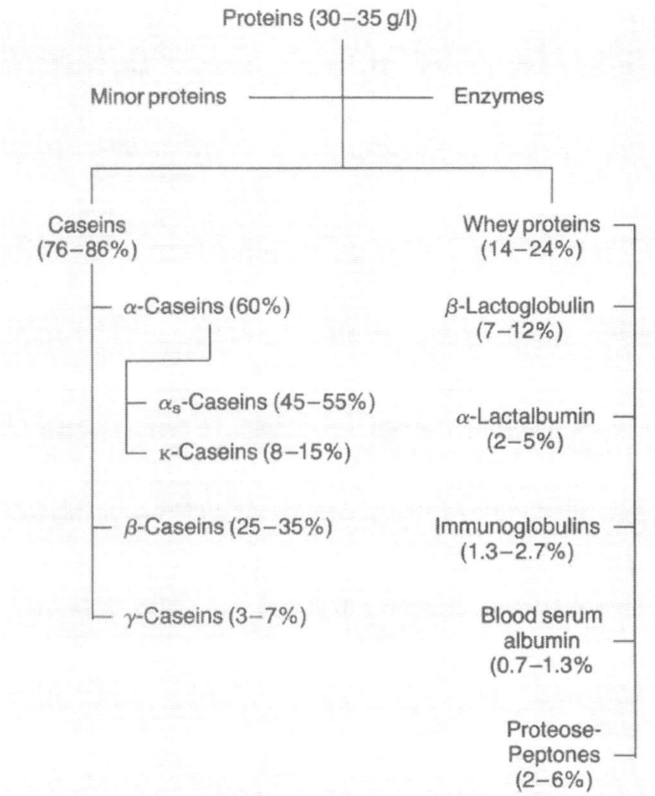

Fig. *Milk protein fractions.*

Casein is easily separated from milk, either by acid precipitation or by adding rennin. In cheese-making most of the casein is recovered with the milk fat. Casein can also be recovered from skim milk as a separate product.

Casein is dispersed in milk in the form of micelles. The micelles are stabilised by the Ê-casein. Caseins are hydrophobic but Ê-casein contains a hydrophilic portion known as the glycomacropeptide and it is this that stabilises the micelles. The structure of the micelles is not fully understood.

When the pH of milk is changed, the acidic or basic groups of the proteins will be neutralised. At the pH at which the positive charge on a protein equals exactly the negative charge, the net total charge of the protein is zero. This pH is called the isoelectric point of the protein (pH 4.6 for casein). If an acid is added to milk, or if acid-producing bacteria are allowed to grow in milk, the pH falls. As the pH falls the charge on casein falls and it precipitates. Hence milk curdles as it sours, or the casein precipitates more completely at low pH.

WHEY PROTEINS

After the fat and casein have been removed from milk, one is left with whey, which contains the soluble milk salts, milk sugar and the remainder of the milk proteins. Like the proteins in eggs, whey proteins can be coagulated by heat. When coagulated, they can be recovered with caseins in the manufacture of acid-type cheeses. The whey proteins are made up of a number of distinct proteins, the most important of which are b-lactoglobulin and lactoglobulin. b-lactoglobulin accounts for about 50% of the whey proteins, and has a high content of essential amino acids. It forms a complex with Ê-casein when milk is heated to more than 75°C, and this complex affects the functional properties of milk. Denaturation of b-lactoglobulin causes the cooked flavour of heated milk.

OTHER MILK PROTEINS

In addition to the major protein fractions outlined, milk contains a number of enzymes. The main enzymes present are lipases, which cause rancidity, particularly in homogenised milk, and phosphatase enzymes, which catalyse the hydrolysis of organic phosphates. Measuring the inactivation of alkaline phosphatase is a method of testing the effectiveness of pasteurisation of milk.

Peroxidase enzymes, which catalyse the breakdown of hydrogen peroxide to water and oxygen, are also present. Lactoperoxidase can be activated and use is made of this for milk preservation.

Milk also contains protease enzymes, which catalyse the hydrolysis of proteins, and lactalbumin, bovine serum albumin, the immune globulins and lactoferrin, which protect the young calf against infection.

MILK CARBOHYDRATES

Lactose is the major carbohydrate fraction in milk. It is made up of two sugars, glucose and galactose. The average lactose content of milk varies between 4.7 and 4.9%, though milk from individual cows may vary more. Mastitis reduces lactose secretion.

Lactose is a source of energy for the young calf, and provides 4 calories/g of lactose metabolised. It is less soluble in water than sucrose and is also less sweet. It can be broken down to glucose and galactose by bacteria that have the enzyme b-galactosidase. The glucose and galactose can then be fermented to lactic acid. This occurs when milk goes sour. Under controlled conditions they can also be fermented to other acids to give a desired flavour, such as propionic acid fermentation in Swiss-cheese manufacture.

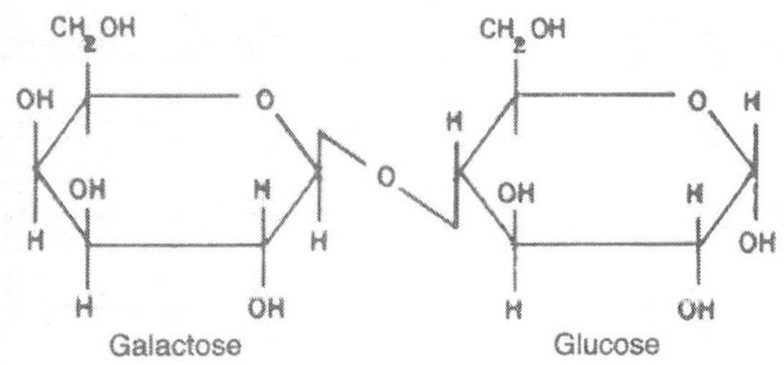

Fig. *Structure of a lactose molecule.*

Lactose is present in milk in molecular solution. In cheese-making lactose remains in the whey fraction. It has been recovered from whey for use in the pharmaceutical industry, where its low solubility in water makes it suitable for coating tablets. It is used to fortify baby-food formula. Lactose can be sprayed on silage to increase the rate of acid development in silage fermentation. It can be converted into ethanol using certain strains of yeast, and the yeast biomass recovered and used as animal feed. However, these processes are expensive and a large throughput is necessary for them to be profitable. For smallholders, whey is best used as a food without any further processing.

Heating milk to above 100°C causes lactose to combine irreversibly with the milk proteins. This reduces the nutritional value of the milk and also turns it brown. Because lactose is not as soluble in water as sucrose, adding sucrose to milk forces lactose out of solution and it crystallises. This causes sandiness in such products as ice cream. Special processing is required to crystallise lactose when manufacturing products such as instant skim milk powders.

Some people are unable to metabolise lactose and suffer from an allergy as a result. Pre-treatment of milk with lactase enzyme breaks down the lactose and helps overcome this difficulty. In addition to lactose, milk contains traces of glucose and galactose. Carbohydrates are also present in association with protein. Ê-casein, which stabilises the casein system, is a carbohydrate-containing protein.

NUTRITIONAL COMPONENTS IN MILK

This page describes the function of nutritional components in milk: Energy,Water, Carbohydrate, Fat, Protein, Vitamins, Minerals, and Minor Biological Proteins & Enzymes. Links are provided to move the reader to pages that present the content of specific nutrients in milk, important background information on the chemistry of milk carbohydrate (lactose), fat, protein, and enzymes.

Energy

The energy in milk comes from its protein, carbohydrate and fat content, with the exception of skim milk that has virtually no fat. The energy content of some milk varieties is shown in the Nutrient Content Tables. Food provides energy to the body in the form of calories (kcal). There are many components in food that provide nutritional benefits, but only the macronutrients protein, carbohydrate and fat provide energy. The energy value of a food is calculated based on the calories provided by the amount of protein (4 kcal/gram), carbohydrate (4 kcal/gram), and fat (9 kcal/gram) that is present.

Water

Milk is approximately 87% water, so it is a good source of water in the diet. Thewater content of some milk varieties is shown in the Nutrient Content Tables.

Water does not provide a nutritional benefit in the same manner as proteins or vitamins, for example. However, water is extremely important in human metabolism. Water is a major component in the body. Water maintains blood volume, transports nutrients like glucose and oxygen to the tissues and organs, and transports waste products away from tissues and organs for elimination by the body. Water helps to lubricate joints and cushions organs during movement. Water maintains body temperature regulation through sweating. Lack of water (dehydration) results in fatigue, mental impairment, cramping, and decreased athletic performance. Severe dehydration can be life-threatening.

Carbohydrate

Milk is approximately 4.9% carbohydrate in the form of lactose. The lactosecontent of some milk varieties is shown in the Nutrient Content Tables.

Carbohydrates are the primary source of energy for activity. Glucose is the only form of energy that can be used by the brain. Excess glucose is stored in the form of glycogen in the muscles and liver for later use. Carbohydrates are important in hormonal regulation in the body. Lack of adequate levels of glucose in the blood and carbohydrate stores leads to muscle fatigue and lack of concentration.

Lactose is a disaccharide made up of glucose and galactose bonded together. Before it can be used by the body, the bond must be broken by the enzyme lactase in the small intestine. People that have decreased activity of lactase in the small intestine may have problems digesting lactose and this is referred to as lactose intolerance or malabsorption.

Fat

Milk is approximately 3.4% fat. The fat content of some milk varieties is shown in the Nutrient Content Tables.

Fats are a structural component of cell membranes and hormones. Fats are a concentrated energy source and are the main energy source used by the body during low intensity activities and prolonged exercise over 90 minutes. Fat is the main storage form of excess energy in the body. Fats cushion organs during movement. There are 2 fatty acids that are considered "essential" that cannot be made by the body and must come from the diet, and these are linoleic (18:2) and linolenic (18:3) acids. These fatty acids are used to synthesize the longer chain fatty acids arachidonic acid (AA, 20:4o-6), docopentaenoic acid (DPA, 22:4o-6), eicosapentaenoic acid (EPA, 20:5o-3) and docohexaenoic acid (DHA, 22:6o-3). These fatty acids are essential for the synthesis of hormones such as prostaglandins, thromboxanes, and leukotrienes that are involved in muscle contraction, blood clotting, and immune response.

The fatty acids in milk fat are approximately 65% saturated, 29% monounsaturated, and 6% polyunsaturated. The polyunsaturated fatty acids in milk fat include small amounts of the essential fatty acids linoleic and linolenic, and approximately 5% *trans* fatty acids. An important *trans* fatty acid in milk fat is conjugated linoleic acid (CLA, 18:2). There are several types (isomers) of CLA in milk that have been shown to inhibit cancer and help maintain lean body mass while promoting the loss of body fat.

The health concerns associated with fats are often linked to the chemical differences in the fatty acids. Saturated and *trans* unsaturated fats have been associated with high blood cholesterol and heart disease. However, the relationships are not simple. The length of the fatty acid chain and source of the unsaturated bond (naturally-occurring or man-made through processing) can greatly influence the health consequences of a specific fat in the human diet. In addition, the genetics and health status of an individual greatly influences the impact of consuming different types of fats. The subject of fats and health is complex and constantly being updated in the medical literature. Issues relating to milk fat.

The content of cholesterol in milk is shown in the Nutrient Content Tables. Cholesterol is an important component of cell membranes and as a starting material for the production of bile salts and steroid hormones. The body manufactures cholesterol to ensure that an adequate level of cholesterol is available for body functions. High levels of blood cholesterol are associated with increased risk for heart disease and are discussed in the Milk and Human Health section. Cholesterol is associated with fat so the content will vary depending on the fat content of the dairy product. An 8 oz serving of 2% milk contains 8% of the Daily Reference Intake (DRI) for cholesterol.

Protein

Milk is approximately 3.3% protein and contains all of the essential amino acids. The protein content of some milk varieties is shown in the Nutrient Content Tables.

Proteins are the fundamental building blocks of muscles, skin, hair, and cellular components. Proteins are needed to help muscles contract and relax, and help repair damaged tissues. They play a critical role in many body functions as enzymes, hormones, and antibodies. Proteins may also be used as an energy source by the body.

Nine amino acids must be obtained from the diet and are referred to as the "essential" amino acids: leucine, isoleucine, valine, phenylalanine, tryptophan, histidine, threonine, methionine, and lysine. Proteins that contain all 9 essential amino acids are often called "complete" proteins. Proteins of animal origin and soy are complete proteins, whereas proteins from grains and legumes are missing 1 or more of the essential amino acids, which means that consumers must eat complementary foods in order to get all of the essential amino acids.

Milk protein consists of approximately 82% casein and 18% whey (serum) proteins. Both casein and whey proteins are present in milk, yogurt, and ice cream. In most cheeses the casein is coagulated to form the curd, and the whey is drained leaving only a small amount of whey proteins in the cheese. During cheese making, the 6-casein is cleaved between specific amino acids and results in a unique protein fragment that is drained with the whey. This fragment, called milk glycomacropeptide, does not have any phenylalanine and can be used as a source of protein for people with phenylketonuria, the inability to digest proteins that contain phenylalanine. Whey proteins have become popular ingredients in foods as an additional source of protein or for functional benefits. Whey proteins are used as a protein source in high protein beverages and energy bars targeted to athletes. Some examples include the use of whey proteins to bind water in meat and sausage products, provide a brown crust in bakery products, and provide whipping properties that replace a portion of egg whites.

Whey proteins contain immunoglobulins which are important in the immune responses of the body. Whey proteins contain branched chain amino acids (leucine, isoleucine, and valine) and have been proposed to have some benefits to athletes for muscle recovery and for preventing mental fatigue.

Vitamins

Vitamins have many roles in the body including metabolism co-factors, oxygen transport and antioxidants. They help the body use carbohydrates, protein, and fat. The functions of vitamins are described below in alphabetical order.

The content of vitamin A in milk is shown in the Nutrient Content Tables. Vitamin A is a fat soluble vitamin involved in vision, gene expression, reproduction, and immune response. The compounds with vitamin A activity are called retinoids and are found in foods in different forms – typically animal foods provide retinol and retinyl esters, and plant foods provide ß-carotene, a starting molecule (precursor) for vitamin A synthesis. Milk contains retinol, retinyl esters, and ß-carotene. Dairy products are a good source of vitamin A, although the vitamin A content will vary with the fat content of the product. An 8 oz serving of 2% milk contains approximately 15% of the daily reference intake

(DRI) for vitamin A. The content of thiamin (vitamin B1) in milk is shown in the Nutrient Content Tables. Thiamin is a water soluble vitamin that is an enzyme cofactor involved in the metabolism of carbohydrates and branched chain amino acids. An 8 oz serving of 2% milk contains approximately 8% of the DRI for thiamin.

The content of riboflavin (vitamin B2) in milk is shown in the Nutrient Content Tables. Riboflavin is a water soluble vitamin that is an enzyme cofactor involved in electron transport reactions. Milk is a recommended source of riboflavin and an 8 oz serving of 2% milk provides approximately 35% of the DRI for riboflavin.

The content of niacin (vitamin B3) in milk is shown in the Nutrient Content Tables. Niacin is a water soluble vitamin that is an enzyme cofactor involved in electron transport reactions required for energy metabolism. There is a small amount of niacin in milk, an 8 oz serving of 2% milk contains less than 2% of the DRI for niacin.

The content of pantothenic acid (vitamin B5) in milk is shown in the Nutrient Content Tables. Pantothenic acid is a water soluble vitamin that is an enzyme cofactor in fatty acid metabolism. Milk is a good source of pantothenic acid and an 8 oz serving of 2% milk contains approximately 17% of the DRI for pantothenic acid.

The content of vitamin B6 (pyridoxine) in milk is shown in the Nutrient Content Tables. Vitamin B6 is a water soluble vitamin involved in the metabolism of proteins and glycogen (energy stored in the liver and muscles), and in the metabolism of sphingolipids in the nervous system. An 8 oz serving of 2% milk contains approximately 7% of the DRI for vitamin B6.

The content of vitamin B12 (cobalamin) in milk is shown in the Nutrient Content Tables. Vitamin B12 is a water soluble vitamin involved in protein metabolism and blood functions. Milk is a recommended source of vitamin B12. An 8 oz serving of 2% milk contains approximately 47% of the DRI for vitamin B12.

The content of vitamin C in milk is shown in the Nutrient Content Tables. Vitamin C is a water soluble vitamin that is an important antioxidant. It has a role in collagen formation in connective tissue and helps in iron absorption and healing of wounds and injuries. There is a negligible amount of vitamin C in milk, and a serving of milk contains less than 1% of the DRI for Vitamin C.

The content of vitamin D in milk is shown in the Nutrient Content Tables. Vitamin D is a fat soluble vitamin that is important in maintaining blood calcium and phosphorus balance and assists calcium metabolism. Milk is typically fortified with vitamin D. Fortified milk is a good source of vitamin D, and an 8 oz serving of 2% milk contains over 50% of the DRI for vitamin D.

The content of vitamin E in milk is shown in the Nutrient Content Tables. Vitamin E is a fat soluble vitamin that has antioxidant activity. The compounds with vitamin E activity are the tocopherols and tocotrienols. Milk contains a small amount of vitamin E, which increases with increasing fat content of dairy products. An 8 oz serving of whole milk contains 1% vitamin E, and an 8 oz serving of 2% milk contains only 0.5% of the DRI for vitamin E.

The content of folate in milk is shown in the Nutrient Content Tables. Folate is one of the water soluble B vitamins. Folate is an enzyme cofactor important in the metabolism of proteins and nucleic acids and blood functions. There is a small amount of folate in milk. An 8 oz serving of 2% milk contains 3% of the DRI for folate.

The content of vitamin K in milk is shown in the Nutrient Content Tables. Vitamin K is a fat soluble vitamin involved in blood clotting, bone metabolism, and protein synthesis. Milk contains a small amount of vitamin K, which increases with the fat content in dairy products. An 8 oz serving of milk contains less than 1% of the DRI for vitamin K.

Minerals

Minerals have many roles in the body including enzyme functions, bone formation, water balance maintenance, and oxygen transport. They help the body use carbohydrates, protein, and fat. The functions of minerals are described below in alphabetical order.

The content of calcium in milk is shown in the Nutrient Content Tables. Calcium plays an essential role in bone formation and metabolism, muscle contraction, nerve transmission and blood clotting. Dairy products are a significant source of calcium in the diet. Milk is a recommended source of calcium, and an 8 oz serving contains almost 30% of the DRI for calcium.

The content of copper in milk is shown in the Nutrient Content Tables. Copper is a component of enzymes used in iron metabolism. Milk contains a small amount of copper. An 8 oz serving of 2% milk contains approximately 3% of the DRI for copper.

The content of iron in milk is shown in the Nutrient Content Tables. Iron is a component of blood and many enzymes. It is involved in blood metabolism and oxygen transport. Milk contains a small amount of iron, and an 8 oz serving of milk contains less than 1% of the DRI for iron.

The content of magnesium in milk is shown in the Nutrient Content Tables. Magnesium is an enzyme cofactor and is important in bone metabolism. Milk is a recommended source of magnesium, and an 8 oz serving of 2% milk contains approximately 7% of the DRI for magnesium.

The content of manganese in milk is shown in the Nutrient Content Tables. Manganese is involved in bone formation, and in enzymes involved in amino acid, cholesterol, and carbohydrate metabolism. There is a small amount of manganese in milk. An 8 oz serving contains less than 1% of the DRI.

The content of phosphorus in milk is shown in the Nutrient Content Tables. Phosphorus is involved in maintaining body pH, in storage and transfer of energy, and in nucleotide synthesis. Milk is a recommended source of phosphorus, and an 8 oz serving of milk contains over 30% of the DRI for phosphorus. The content of potassium in milk is shown in the Nutrient Content Tables. Potassium is an electrolyte that is important in the maintenance of water balance, blood volume and blood pressure. Dairy products are a recommended source of potassium, and an 8 oz serving of milk contains approximately 8% of the DRI for potassium.

The content of selenium in milk is shown in the Nutrient Content Tables. Selenium is important in oxidative stress response, electron transport, and regulation of thyroid hormone. Milk is a good source of selenium, and an 8 oz serving of 2% milk contains approximately 11% of the DRI for selenium.

The content of sodium in milk is shown in the Nutrient Content Tables. Sodium is an electrolyte that is important in the maintenance of water balance and blood volume. An 8 oz serving of milk contains approximately 7% of the DRI for sodium.

The content of zinc in milk is shown in the Nutrient Content Tables. Zinc is a component of many enzymes and proteins, and is involved in gene regulation. Milk is a good source of zinc, and an 8 oz serving contains approximately 10% of the DRI for zinc.

Minor Biological Proteins & Enzymes

Other minor proteins and enzymes in milk that are of nutritional interest include lactoferrin and lactoperoxidase. There are many other enzymes in milk but these do not have a role in human nutrition.

Lactoferrin is an iron binding protein that plays a role in iron absorption and immune response. Many other functions of lactoferrin have been proposed, but their confirmation is still under study, including protection against bacterial and viral infections, and it's role in inflammatory response and enzyme activity.

Lactoperoxidase is an enzyme that, in the presence of hydrogen peroxide and thiocyanate, has antibacterial properties. The use of lactoperoxidase as an antimicrobial agent is discussed in the section on Antibacterial Properties of Milk in this website. Lactoperoxidase does not provide antimicrobial protection to fresh milk because hydrogen peroxide is not normally present in milk – it must be added to activate this system.

Lipases, a group of enzymes that break down fats, are present in milk but are inactivated by pasteurization, which increases the shelf life of milk. A popular belief among raw milk consumers is that the native lipase in milk plays an important role in the digestion of fat. Fat digestion begins in the stomach with gastric lipase, and the majority of fat digestion occurs in the small intestine, using enzymes secreted by the pancreas. The relative importance of the native milk lipase in digestion compared to the pancreatic lipases is not clear. Lactase (ß-galactosidase) is the enzyme responsible for the breakdown of lactose into glucose and galactose for digestion. There is no lactase present in fresh milk.

Any lactase present in milk products comes from lactic acid bacteria that are either added to milk on purpose, as in the case of yogurt and cheese, or that enter milk from airborne or other contamination. A popular belief is that people with lactose intolerance are able to drink raw milk but not pasteurized milk because the lactase present in raw milk is inactivated during pasteurization. Because there is no lactase present in fresh milk, this concept is a myth. People with lactose intolerance have, themselves, lower levels of lactase which creates problems when it comes to digesting large amounts of lactose in a timely manner. Naturally occurring lactase used to digest milk is normally secreted by the small intestine. Lactase found in any lactic acid bacteria present will minimally help to digest lactose when it is released as the milk is digested in the small intestine.

MINOR MILK CONSTITUENTS

In addition to the major constituents, milk also contains a number of organic and inorganic compounds in small or trace amounts, some of which affect both the processing and nutritional properties of milk.

MILK SALTS

Milk salts are mainly chlorides, phosphates and citrates of sodium, calcium and magnesium. Although salts comprise less than 1 % of the milk they influence its rate of coagulation and other functional properties. Some salts are present in true solution. The physical state of other salts is not fully understood. Calcium, magnesium, phosphorous and citrate are distributed between the soluble and colloidal phases. Their equilibria are altered by heating, cooling and by a change in pH.

Table. ***Distribution of milk salts between the soluble and colloidal phases.***

	Total	Dissolved	Colloidal
	(mg/100 ml of milk)		
Calcium	1320.1	51.8	80.3
Magnesium	10.8	7.9	2.9
Total phosphorus	95.8	36.3	59.6
Citrate	156.6	141.6	15.0

In addition to the major salts, milk also contains trace elements. Some elements come to the milk from feeds, but milking utensils and equipment are important sources of such elements as copper, iron, nickel and zinc.

MILK VITAMINS

Milk contains the fat-soluble vitamins A, D, E and K in association with the fat fraction and water-soluble vitamins B complex and C in association with the water phase. Vitamins are unstable and processing can therefore reduce the effective vitamin content of milk.

SEPARATION OF MILK COMPONENTS

CREAM

When milk is left to stand for some time, fat globules rise to the surface forming a layer of fat (or cream). This can be separated leaving behind skimmed milk as a by-product.

There are different types of cream each with different fat concentrations: single (or light) cream contains 18 per cent milk fat whereas double (or heavy) cream normally contains 30 per cent milk fat. Cream is a luxury item and may be used as an accompaniment to coffee, as a filling in cakes, and an ingredient in ice cream.

Separation

Separation can very simply be achieved by removing the cream with a spoon, however this is a slow process during which the cream may spoil. For this reason it is more usual to use a manual or powered centrifuge.

Table. Production stages for cream

Ingredients	*Process*	*Equipment*
Raw milk tested	Store at 4°C	Milk churns
		Refrigerated storage
		Thermometer
	Separation of milk fat	Ladle Dairy centrifuge
	Pasteurization	Large boiling pan or steam jacketed pan
		Pasteurizer
		Heat source
		Thermometer
	Fill bottles/pots	Funnel or liquid-filling machine
		Capping machine
		Pot sealer
	Cool bottles	Bottle-cooler
	Store bottles at 4°C	Refrigerated storage

Pasteurization

Cream may be pasteurized in a similar way to milk, using a similar time and temperature combination and the same equipment. Cream can also be sterilized but there is a considerable loss of quality.

Packaging and storage

Cream can be packaged in glass jars or plastic pots sealed with foil lids. Pasteurized cream must be stored at a temperature of 4°C to have a shelf-life of several days. Refrigerated storage is necessary because cream is prone to rapid spoilage.

BUTTER

Butter is a semi-solid mass which contains approximately 80-85 per cent milk-fat and 15-20 per cent water. It is yellow/white in colour, with a bland flavour and a slightly salty taste.

It is a valuable product that has a high demand for domestic use in some countries and as an ingredient in other food processing (e.g. for confectionery and bakery uses).

The principles of preservation are:

- To destroy enzymes and micro-organisms by pasteurizing the milk
- To prevent microbial growth during storage by reducing the water content, by storing the product at a low temperature, and optionally by adding a small amount of salt during processing.

Table. Production stages for butter

Ingredients	*Process*	*Equipment*
Cream or soured cream	Store at 4°C	Milk churns Refrigerated storage Thermometer
	Churning	Butter churns
	Draining (pour off buttermilk)	
	Washing	
	Draining (pour off washwater)	
Permitted colours and salt (optional)	Kneading/working	Butter pats
	Form into blocks	Butter pats
	Packaging	Paper/plastic/ foil wrapping Wrapping machines
	Storage at 4°C	Refrigerated storage

Churning

Churning disrupts the emulsion of fat and water and as a result the milk-fat separates out into granules. This process takes place in a butter churn.

Churning cream

Churning is continued until fat granules are present and at this stage the mixture is drained to remove liquid that has separated from the granules. This liquid is known as buttermilk and can be used as either a beverage or as an ingredient in animal feed.

Washing

Clean water equivalent in weight to the buttermilk is added to the churn in order to wash the butter granules.

The wash water is drained away. Churning is continued for a short time to compact the butter, and once this has been achieved it is removed from the churn.

Forming and packaging

Butter is kneaded to achieve a smooth and pliable texture. This can be done using simple hand-tools such as butter pats. Alternatively for higher production rates a specially-designed kneader can be used. Once the butter has a uniform and smooth texture it is formed into blocks with butter pats and packed in either greaseproof paper or foil wrappers.

WORKING BUTTER WITH BUTTER PATS

Storage

Due to its high fat composition, butter must be stored at temperatures below 10°C otherwise the fat becomes rancid and imparts undesirable 'off' flavours. The water droplets in butter (20 per cent) can also allow bacteria to grow if it is not kept under cool conditions.

Ghee

Ghee is made from butter which has been heated and clarified. At ambient temperatures it is a semi-solid mass with a granular texture, but on melting (40°C+) it turns into a clear, thin liquid. It has a high demand in some countries for domestic use, as an ingredient for local food production (for example bakeries and confectionery manufacturers), and as an export commodity.

Alternatively, cream is boiled gently to evaporate the water. During boiling the product is stirred continuously until the milk proteins start to coagulate, forming particles, and the colour of the cream darkens. Heating is stopped and the product is left to set. The particles settle at the bottom of the vessel and the milk-fat is separated. The principles of preservation are:

- Heating to destroy enzymes and contaminating microorganisms
- To reduce the water-content by evaporation, and in doing so prevent the growth of micro-organisms.

Table. Production stages for ghee

Ingredients	*Process*	*Equipment*
Butter	Heating	Heat source
		Large boiling pan or steam jacketed pan
	Cool to room temperature	Thermometer
	Filter	Filter cloth
	Fill into jars/pots	Funnel or liquid-filling machine
		Capping machine
	Store at ambient temperatures	

Packaging and storage

Metal containers are normally used. They should be thoroughly cleaned, especially if they are re-usable, and they should be made airtight. Alternatives to metal cans include coloured glass jars with metal lids, or ceramic pots sealed with cork/plastic stoppers.

Ghee is usually stored at room temperatures as cold storage affects the granular texture. Thus ghee is useful for those consumers with no access to refrigeration.

CULTURED/FERMENTED DAIRY PRODUCTS

The technology of cultured milk products such as yoghurt, curd, and cheese is based upon the microbial conversion of the milk-sugar lactose to lactic acid (lactic acid accounts for the characteristic 'sourness' of such products). In order for the conversion to take place, lactic acid producing bacteria must be present. This may occur by allowing the milk to sour naturally, but it is better to introduce the appropriate bacteria as a starter culture. Starter cultures may be in the form of a small quantity of previously-cultured product or may be purchased as a commercially-prepared culture.

Yoghurt/curd

Yoghurt is a fermented milk product that evolved by allowing naturally-contaminated milk to sour at a warm temperature. Yoghurt can be either unsweetened or sweetened, set, or stirred. Curd is the name given to a yoghurt-type product made from buffalo milk.

The principles of preservation for yoghurt are:

- Pasteurization of the raw milk to destroy contaminating microorganisms and enzymes.
- An increase in acidity due to the production of lactic acid from lactose. This inhibits the growth of food-poisoning bacteria.
- Storage at a low temperature to inhibit the growth of microorganisms.

Table. Production stages for set yoghurt

Ingredients	*Process*	*Equipment*
Milk and starter culture (2 per cent)	Preheat to 70°C for 15-20 minutes	Heat source
		Thermometer
		Boiling pan
	Cool to 30-40°C	Thermometer
	Addition of starter culture	Measuring and weighing equipment
	Pour into bottles/pots	Funnel or Liquid filler
		Sealing machine
		or Capping machine
	Incubate at 43-45°C	Commercial incubator
		Thermometer
	Store at 4°C	Refrigerated storage

Heating

In the manufacture of yoghurt, milk is normally heated to 70°C for 15-20 minutes, using an open boiling pan, or alternatively a steam jacketed pan.

Addition of starter culture

The milk is cooled to between 30 40°C and inoculated with a mixed culture of *Lactobacillus bulgaricus* and *Streptococcus thermophilus* (usually in a ratio of 1:1). If a commercial starter-culture is used, the directions for use will be given. However, if a culture from a previous batch is used, then it is usual to add 2-3 tablespoons per litre of prepared milk.

Yoghurt of the stirred variety can be fermented in the mixing container. To make set yoghurt the inoculated milk should be poured into the individual pots before fermentation.

SELLING CURD FROM A ROADSIDE STALL

Incubation

The micro-organisms that produce yoghurt are most active within a temperature range of 32-47°C. Ambient temperatures are therefore not adequate and a heated incubator is needed. Small commercially-available yoghurt-makers consist of an electrically-heated base and a set of plastic or glass containers.

Most yoghurt-makers make four or five individual half litre cups at a time. There are other simple and inexpensive ways of incubating yoghurt such as an insulated box, keeping the jars/pots surrounded by warm water, or by using thermos flasks (the latter is only suitable for stirred yoghurt). Incubation takes approximately five hours.

When fermentation is complete, stirred yoghurt is cooled and flavoured or sweetened prior to packaging. In set yoghurt all ingredients are added before fermentation.

Packaging and storage

Yoghurt or curd is commonly packaged in plastic pots fitted with a plastic lid, or heat-sealed with foil, although traditionally, curd is packaged in clay pots. Such pots are made from local materials and can be re-used or later used for cooking. The shelf-life of yoghurt is usually 3-8 days when stored at temperatures below 10°C.

Cheese

Cheese is made from milk by the combined action of lactic acid bacteria and the enzyme rennin (known as rennet). Just as cream is a concentrated form of milk fat, cheese is a concentrated form of milk-protein.

The differences in cheeses that are produced in different regions result from variations in the composition and type of milk, variations in the process, and the bacteria used. The different cheese varieties can be classified as either hard or soft.

COLLECTING MILK FOR CHEESE-MAKING

Hard cheeses such as Cheddar and Edam have most of the whey drained out and are pressed. Soft cheeses such as paneer contain some of the whey and are not pressed. Many indigenous cheeses are soft types. The hardness, flavour, and other qualities of a cheese can be varied by changes to the process conditions, to suit local tastes. However the principal steps of a cheese-making process are basically the same.

The principles of preservation are:

- the raw milk is pasteurized to destroy most enzymes and contaminating bacteria
- fermentation by lactic-acid bacteria increases the acidity which inhibits the growth of food-poisoning and spoilage bacteria
- the moisture content is lowered and salt is added to inhibit bacterial and mould growth.

The table, right, outlines the stages of production and the equipment needed to produce Edam cheese.

Pre-heating

The pasteurized milk is heated to a temperature at which the starter-culture can work.

Addition of starter culture

Starter-culture is added to the milk at the rate of approximately 2 per cent of the weight of milk. The vessel used should be either aluminium or stainless steel.

Addition of rennet

The rennet should be 1 per cent of the weight of milk. The rennet alters the milk proteins and allows them to form the characteristic curd.

Incubation

The milk is allowed to stand until it sets to a firm curd.

Treatment of the curd

The curd is cut into cubes which facilitate the elimination of whey from the gel. The curd is then cooked at 40°C for a period of twenty minutes which has the action of firming the curd. After cooling, the whey is drained off. The curd is pressed to ensure that most of the whey has been removed, and is then cut to fit the cheese-moulds, and finally pressed with weights.

Ripening

This is the final stage in the cheese-making process. It is a process which allows the development of gas in some cheeses and the development of flavour.

The longer the ripening process the stronger the flavour. Ripening usually takes place in ripening rooms, where the temperature and humidity must be controlled for the optimum development of the cheese.

Table. Production stages for Edam-type cheese

Ingredients	*Process*	*Equipment*
Pasteurized milk	Preheat to 35-40°C	Cheese vat
		or boiling pan
		Thermometer
		Heat source
Starter culture	Addition of starter culture	Measuring and weighing equipment
Rennet	Add rennet at 30°C	Measuring and weighing equipment
		Thermometer
	Incubate	
	Cut the curd	Curd cutters
	Heat to 40°C for 20 minutes	Heat source
		Thermometer
	Drain	Filter cloth
	Cut to fit a cheese-mould	Knife
	Put into a cheese-mould	Cheese-moulds
	Press with	Cheese-press weights
	Cool and dry at 10-12°C	Thermometer
Salt	Salting in 20 per cent salt solution at 12°C for 12-16 hours	Brine meter
		Thermometer
	Ripen for 6-8 weeks at 16°C	Thermometer (optional)
	Washing	
	Drying for 30 minutes	
	Wax with paraffin wax store at 9°C	Refrigerated storage

Packaging and storage

The packaging requirements differ according to the type of cheese produced. Hard cheese, for example, has an outer protective rind which protects the cheese from air, microorganisms, light, moisture-loss or pick-up, and odour pickup. Cheese should be allowed to 'breathe', otherwise it will sweat. Suitable wrapping materials are therefore cheesecloth or grease-proof paper. Cheese should be stored at a relatively low temperature between 4 and 10°C to achieve a shelf-life of several weeks/months. Soft cheeses are often stored in pots or other containers, often in brine, to help increase their shelf-life of several days/weeks.

SELLING ICE CREAM FROM A BICYCLE

Ice cream

Ice cream is a frozen mixture which contains milk, sugar, fat, and optional thickeners (e.g. pectin or gelatin), colouring, and flavouring. It may be sweetened and flavoured in numerous ways with nuts, fruit pieces, and natural or artificial flavours and colours. The principles of preservation are:

- Pasteurization to destroy most micro-organisms and enzymes
- Freezing to inhibit microbial growth.

Pasteurization

Pasteurization is carried out by heating to 65°C for a period of 30 minutes.

Cooling and beating

Ice cream is a complex mix of small ice crystals and air bubbles in a milk-fat/water emulsion. To achieve this, it is necessary to cool the mixture quickly to produce small ice crystals and at the same time incorporate air into the product by beating.

Ingredients	*Processing stage*	*Equipment*
Milk, sugar, fat, thickener, colours and flavours.	Mix ingredients	Weighing and measuring equipment
		Liquid mixer
	Pasteurize	Boiling pan or steam jacketed pan
		Thermometer
		Heat source
	Cool mixture to approximately -5°C and beat simultaneously	Ice cream maker
	Fill into containers	Filling machine
	Freeze at -18°C	Freezer

Ice cream makers are available commercially and work on the following principle. The mixture is placed in a bowl which is kept at a low temperature (either surrounded by ice and salt, or having been chilled in a freezer). It is then agitated by a manually-operated rotor or by a powered stirrer. At the end of this process the ice cream should be at a temperature of approximately -5°C, and be partly frozen.

Packaging and storage

The ice cream is usually packaged in plastic, waxed paper, or cardboard containers, and is stored at below -18°C. The storage temperature is important for two reasons:

- To maintain the texture of the product
- To prevent the growth of micro-organisms.

Ice cream may be transported in an insulated box (e.g. for sale from a bicycle). It is especially important to guard against thawing and re-freezing as this will cause changes in texture and mouthfeel, and there is the increased possibility of food poisoning by contaminating food poisoning microorganisms.

SUITABILITY FOR SMALL-SCALE PRODUCTION

In some regions, there is a high demand for dairy products both for traditional and modern items. Much of the technology and machinery necessary for processing is fairly simple, which may at first sight appeal to the would-be processor. Milk, however, is a highly perishable food and there is a high risk of transmitting food-poisoning bacteria to consumers. It is stressed, therefore, that milk processing of any kind must be done under carefully controlled hygienic conditions.

Furthermore, in developing countries, milk processing is often more problematic than in temperate climates, owing to higher temperatures and humidity. Consequently, milk spoils at a faster rate, cheese ripens too quickly, and it is often difficult to ensure adequate cooling conditions.

MILK CREAM SEPARATOR

Objective: To study the parts of mechanical cram separator. To assembling and dismantling of the cream separator.

SEPARATOR (MILK)

A separator is a centrifugal device that separates milk into cream and skimmed milk. Separation was commonly performed on farms in the past. Most farmers milked a few cows, usually by hand, and separated milk. Some of the skimmed milk was consumed while the rest was used to feed calves and pigs. Enough cream was saved to make butter, and the excess was sold.

Today, milk is separated in industrial dairies. Sufficient cream is returned to the skimmed milk before sale.

History

Before the advent of centrifugal separators, separation was performed by letting milk sit in a container until the cream floated to the top and could be skimmed off by hand. A variant container-separator had a nozzle at the bottom which was opened to allow the milk to drain off. A window in the side, near the nozzle at the bottom, allowed the operator to observe when the milk was drained.

The centrifugal separator was first manufactured by Gustaf de Laval, making it possible to separate cream from milk faster and more easily, without having to let the milk sit for a time, and risk it turning sour. Possibly because Gustaf de Laval manufactured the first cream separators, many people credit the invention to de Laval. However, many patents appear before his, all of them labelled as 'improvements'. One of the first specifically for cream separation was patented by W. C. L. Lefeldt and C. G. O. Lentsch.

The original centrifugal separators were hand-cranked, as illustrated.

Mechanism

Manual rotation of the separator handle turns a worm gear mechanism which causes the separator bowl to spin at thousands ofrevolutions per minute.

When spun, the heavier milk is pulled outward against the walls of the separator and the cream, which is lighter, collects in the middle. The cream and milk then flow out of separate spouts. Some floor model separators were built with a swinging platform attached to the stand. The bucket for collecting the cream was put on the platform, and a much larger bucket was set on the floor to collect the milk. Some floor model separators had two swinging platforms. Smaller versions of separators were called table-top models, for small dairies with only a few cows or goats.

Gustaf de Laval's construction made it possible to start the largest separator factory in the world, Alfa Laval AB. The milk separator represented a significant industrial breakthrough in Sweden. Within the first decade of the 1900s, there were over twenty separator manufacturers in Stockholm. Separators in modified form are also used on ships to purify oil, which may have been their original use, because in its original form de Laval proposed the separator for use in his steam turbine. De Laval's turbine used mechanically lubricated journal bearings which weren't insulated from the inside of the turbine. When the steam condensed into water it contaminated the oil. To purify the oil a centrifugal separator was used, which was later adapted to the dairy industry.

The original design had a manual bowl that required manual cleaning. Most modern separators use a self-ejecting centrifuge bowl that can automatically discharge any sedimentary solids that may be present, and that allow for clean-in-place (CIP). The self-draining vessel feature is mandatory for machinery designated for 3A approval (an industrial sanitary standard).

Resources

The North American Dairy Foundation publishes a separator and other dairy antique newsletter three times yearly.

RELEVANT INFORMATION

Cream separation is a phenomenon by which the milk is separated in to cream and skim milk by centrifugal and gravitation force. The centrifugal force is thousand times greater than gravitational force. The milk enters the rapidly revolving bowl of the separator. It is then acted upon by gravitational and centrifugal force. The inflow of milk is channelised to the outer wall of the bowl and fills it from outside towards the centre. The centrifugal and gravitational forces keep on continuously and act upon to partition the serum and cream. The milk serum has higher specific gravity (1.036) than that of fat (0.9). Obviously serum is thrown towards the outer periphery and channelised to skim milk out let, while the cream is channelised to the central core and forced out through cream out let, while the cream is channelised to the central core and forced out through cream out let. There are different methods i.e. shallow pan methods, deep pan method, water dilution of cream separation. These methods which have poor efficiency and are used in the absence of mechanical device. Air tight separation is improved by mechanical method of cream separation. Separation occurs in a hermetically sealed system thereby milk is kept out of contact with air with minimum foaming of skim milk; mechanical cream separation is common and is in operation on various scales.

Mechanical Cream Separator

In mechanical cream separation the cream separator is the major machine used, it may be hand operated or power operated. Study of Cream Separator: Construction: The modern centrifugal cream separator consists of

I. A bowl which can be rotated at a high speed (5000-6000rpm) by means of suitable gears and power transmission mechanism.
II. Arrangement for supplying milk to the bowl.
III. Removing the cream and skim milk.
IV) Driving the machine:

Bowl parts:

1. Rubber ring or Gasket: It fits in the bottom of the bowl base and makes the bowl leak proof.
2. Distributor: It distributes the milk into the three notches at the bottom periphery.
3. Bottom disc: It is different from the rest of the disc. It has projections on the outer as well as on inner surface and has three holes. It is also called as separating disc.
4. Intermediate disc: they are 6 to 7 in number. They are similar to the bottom disc expect that these discs do not have projections on inner side. They have projection on the outer periphery only. These discs have three holes. The increased number of discs increases the efficiency of separation.
5. Top disc: This is very important part of the bowl. This is also called as separating or dividing disc. The connection between the skim milk and skim milk outlet is made possible by this disc. The disc has an outlet for the cream in its neck. During separation the cream near the axis of the bowl oozes out from the cream outlet. This disc has screw at its neck. The thickness of cream can be adjusted by adjusting this screw.
6. Bowl Hood: It fits over the top and covers the whole bowl parts.
7. Bowl spanner: Its function is to tighten the bowl nut.

Arrangement for Supplying Milk to the Bowl:

1. Supply can: It holds the milk to be separated.
2. Faucet: It is used for allowing the flow of milk from the supply can into the bowl.

3. Regulator: It regulates the flow of milk into the bowl.
4. Float: It allows only a definite quantity of inflow of milk into the bowl, through the regulator.

Removing of Cream or Skim Milk:

1. Skim milk outlet: The skim milk from the bowl moves between the dividing disc and inside wall of the bowl. At the top, space is provided between to disc and bowl hood, which connects with the skim milk outlet. The skim milk is let out through this cover.
2. Cream outlet: The cream coming out from cream screw located on the top disc is trapped in the cream cover and is let out.

Driving the Machine:

1. Spindle rod: The bowl rests on the spindle rod. The spindle rod has a suitable gear which is engaged with crank gear.
2. Crank handle: The machine is revolved with the crank handle at the speed indicated the manufactures. Generally 60-70 revolutions per minutes are sufficient to develop the speed.

STANDARDIZATION OF MILK

RELEVANT INFORMATION

Standardization of milk refers to the adjustment which means rising or lowering of fat and solids not fat levels of milk. The standardization of milk is commonly done in case of market milk supply and also in case of manufacture of milk products. e.g. condensed milk, milk powder, ice-cream and cheese etc. the standardization is mostly done to have a uniform milk fat content in the finished dairy product. In India, milk is toned to 3 % fat or double toned 1.5 % by standardising the fat content of milk and thus volume of milk is increased so that the milk can be supplied to the consumers at low price and to a large population.

Precaution:

Correct calculations by Pearson's method regarding ingredients to be used for standardization.

Material required:

1. Whole milk,
2. Skim milk/cream
3. Butter oil
4. Skim milk
5. Thermometer

Apparatus:

1. Multipurpose vat.
2. Weighting balance

3. Gerber's fat unit
4. Quevenne's Lactometer with jar
5. Thermometer 6. Richmond's scale

Procedure:

1. Pleasure the quantity of whole milk.
2. Find out the fat SNF and Ts content
3. Measure the quantity of skim milk powder available or cream.
4. Determine the fat % and SNF % of skim milk/skim milk powder/ cream.
5. Decide the fat level or SNF level of the milk to be standardized.
6. Calculate the requirement of skim milk/skim milk powder/cream by person's method.

1. For fat adjustment

B (C-B) or (B-C)

A (A-C) or (C-A)

A = Fat % of cream

B = Fat % of skim milk

C = desired fat %on the final product

(A-C) or (C-A) = Represent the quantity of skim milk

(B-C) or (C-B) = Represent quantity of cream with high fat % B for SNF Adjustment

Y (Y-Z) or (Z-Y)

X (x-Z) or (Z-x)

X = SNF content of whole milk/cream

Y = SNF content of skim milk / powder

Z = Desired level of SNF

(x-Z) N (Z-x) = Represent the required quantity of skim milk

(Y-Z) N (Z-Y) = Represent the required quantity of cream / whole milk.

Weigh the required quantity of skim milk / skim milk powder cream and add in to the vat along with the material of which fat and SNF level is to adjusted.

Mix the contents thoroughly. Take the fat / SNF content of the final product to check their desired level. Sent for further processing and packaging.

Calculations:

1. Calculations how much cream containing 35 % fats to be added to 4000 kgs. Of milk containing 3.5 % fat to raised the fat content of milk to 4 % of fat.
2. Calculate how much scream of 35 % fat and how much milk of 3.5% fat to be mixed to make 400 kgs of milk testing 4 % fat.
3. Calculate how much skim milk should be removed from 300 kg of milk containing 3.5 % fat to increase the fat to 5 % assuming that the separate milk contains negligible amount of fat.

4. Calculate the quantities of 0.1 % fat content should be added to 500 kg of whole milk to 4.5 % fat content to decrease the fat to 3.5 %
5. Calculate the quantities of 0.1 % skim milk and 4.5 % whole milk to require being produce 500 kg of 3.5 % fat standardised milk.
6. Standardised 500 kg of cream of 40 % fat to 20 % fat by adding milk containing 3.0 % fat.

CREAM SKIMMING

Raw milk contains approximately 4% fat. To produce non-fat milk, the cream must first be separated. Whereas warm milk separation takes place at temperatures of 50–60 °C, cold milk separation takes place at temperatures of 4–30 °C. Yet, GEA supports both these processes with the procool system.

The cream skimming separators from the hyvol® proplus line set new standards of product protection and yield maximization.

And, as all our separators are equipped with the hydrosoft system, we guarantee the lowest possible feed pressures, gentle product intake, low shear forces and, as a result, the highest possible separation efficiencies. The hydrosoft system uses no high-maintenance mechanical seals and requires no additional cooling water.

Bacterial Removal

Bacteria, yeasts and mold spores can contaminate milk products, which makes reliable bacterial clarification all the more important. GEA's solutions range from single- and two-stage bacteria removal processes to the special bacterial clarification of milk, the targeted bacterial clarification of cheese milk and a treatment for whey concentrate. Beyond this, separators with the GEA proplus module can be used to recover valuable proteins from solid residues in conjunction with bacteria removal.

Standardization

With the latest standomat system, GEA is able to offer a standardization unit for milk fat to milk protein. The system regularly measures and regulates the defined specifications during the production process. Limit values are precisely maintained to optimize resource use and, as regular sampling is no longer required, laboratory checks can be minimized. The flexible optical process used can be applied to consumer milk, standardized cheese milk, yoghurt products, standardized cream products, recombined milk products as well as condensed milk and coffee cream.

With the dairysmart, GEA now offers a skid that has been designed for hot milk/whey skimming, milk cleaning, cream fat content monitoring and automatic tank standardization. Designed for small and medium-sized dairies, as well as cheese factories, it comprises the ecocream skimming separator for milk and whey in conjunction with an intelligent production kit for smart, flexible product processing and the new IO control system from GEA.

PREPARATION OF CHAKKA AND SHRIKHAND

Shrikhand, a semi-solid, sweetish-sour fermented milk product, is very close to flavoured quarg of Germany. It is a popular dessert of Gujarat, Maharashtra and Karnataka and forms part of meal especially on festive occasions. Shrikhand is prepared by fermentation of milk with lactic acid bacteria, expulsion of whey from the curd, followed by mixing with sugar, flavouring and spices. The composition of this product is: 34-40 percent moisture, 43-45 percent sugar, 4-6 percent fat, and 10-12 percent milk solids-not-fat.

STANDARDS

The standards prescribed for Shrikhand by the Bureau of Indian Standards (BIS) and under the prevention of food adulteration act (PFA) are presented in table .

	BIS	PFA
Total solids (per cent, by mass), minimum	58.0	58.0
Milk fat (in dry matter, per cent by mass), minimum	8.5	8.5
Milk protein (in dry matter, per cent by mass), minimum	10.5	10.5
Titratabel acidity (per cent lactic acid), maximum	1.4	1.4
Sucrose (in dry matter per cent by mass), maximum	72.5	72.5
Total ash (in dry matter per cent by mass), maximum	0.9	0.9
Coliform count, cfu/g, maximum	--	10.0
Yeast and mould count, cfu/g, maximum	--	50.0

Fig. PFA/BIS standards for shrikhand

Chemical Composition

Constituent	Dahi	Chakka	Shrikhand
Total solids (%)	9-10	22-23	57-60
Fat (%)	Trace	Trace	5-6
Proteins (%)	3.7-3.9	13.5-14	6.5-7.0
Sucrose (%)	--	--	40-43
Ash (%)	0.72	0.95-1.08	0.49-0.55
Reducing sugar (%)	4.5-4.7	3-3.25	1.6-1.7
Titratable Acidity (%)	0.9-0.95	2.1-2.2	1.05-1.10
PH	4.4-4.6	4.4-4.6	4.4-4.6

Composition of dahi, chakka, and shrikhand

The chemical composition of shrikhand depend upon the type and quality of milk, processing conditions employed, and fruits/spices used. The physico chemical properties and chemical composition of shrikhand are presented in below tables respectively. The chemical composition of Shrikhand sold in Gujarat is given in table .

Parameters	Average values		
	Maharashtra[1]	Gujarat[1]	Industrial[2]
PH	4.09	4.10	4.33
Acidity (% as lactic acid)	1.30	1.08	1.13
Soluble nitrogen (%)	0.21	0.29	0.29
Free fatty acids (% as oleic acid)	0.58	0.40	0.26
Penetration value (mm) at 10°C	28.54	27.35	32.71

Physico-chemical parameters of shrikhand marketed in Gujarat and Maharashtra

Average values of product made in the un-organized sector.

Average values of product made by the mechanized process in the organized sector.

Constituent	Traditional shrikhand	Industrial shrikhand
Moisture (%)	40.52	35.11
Fat (%) DM*	8.27	9.25
Proteins (%) DM	9.53	12.53
Lactose (%) DM	3.74	2.59
Sucrose (%) DM	76.25	70.00
Ash (%)	0.530	0.683
Titratable acidity (% LA)	1.075	1.310
PH	4.10	4.33
Soluble nitrogen (%)	0.290	0.286
FFA (% oleic acid)	0.395	0.265

Chemical composition of shrikhand sold in Gujarat

Manufacture of Shrikhand

The method of manufacture of shrikhand involves the preparation of curd or dahi by fermentation of milk with starter culture, preparation of chakka by draining whey from the curd and blending additives like sugar, color, flavour, species and fruits to obtain a desired composition and consistency.

(i) Traditional method: The traditional method of making shrikhand consists of preparation of dahi by culturing cow or buffalo milk with natural starter culture (curd from previous batch), preparation of chakka by draining whey from dahi followed by blending sugar, colour, flavouring materials and spices.Milk of cow or buffalo, or mixed milk is heated to boiling and then cooled to 30o-35-oC. It is inoculated with dahi (lactic culture) @ 0.5 to 1 per cent from the previous day's production. Milk is left at room temperature (30o-35oC) until it sets firmly. It is then stirred and hung in a muslin cloth for 10 to 12 hours,to drain off whey. The product so obtained is called chakka or maska. The chakka yield is about 65 kg per 100 kg of milk, depending upon the composition of milk.

The chakka is mixed with the required amount of sugar (usually 50-100 per cent of curd quantity), flavour, colour, herbs and spices. Commonly used additives include saffron, cream, charoli, almonds, nutmeg, cashew nut, mango pulp, raisins, and seasonal fruits. The shrikhand yield is about 1.5 to 2.0 kg per kg of chakka used. Shrikhand is served chilled. Consumers in Gujarat prefer its sweet variety while those in Maharashtra prefer the tangy/soury variety.

(ii) Industrial method: The fully mechanized and continuous process has been developed for industrial production of shrikhand. With the development of the dairy industry, shrikhand production has become industrialized. The process consists of centrifugal separation of whey from the curd and the mechanical mixing of 'chakka;, sugar and species, making the entire process hygienic as well as labour and time saving. The industrial method of shrikhand manufacture.

Normally skim milk is used for making dahi for the manufacture of Shrikhand. By using skim milk, not only are fat losses eliminated, but faster moisture expulsion and less moisture retention in the curd are achieved. Skim milk is heated to 85oC for 30 min. This heat treatment results in denaturation ofâ-lactoglobulin, which in turn interacts with casein on acid coagulation, thereby increasing the yield. Also high temperatures are believed to result in the alteration of proteins in milk, which favours the growth, and activity of starter culture.Also, application of heat to milk results in the destruction of certain heat labile inhibitors as well as many of the competing microorganisms that are present in raw milk. The industrial production involves the following steps:

Preparation of dahi/curd: Skim milk (9% SNF, 0.05% fat) is heated to 90oC for 15 seconds in HTST pasteurizer, cooled to 30oC and inoculated with 0.25 – 0.50 per cent dahi culture of mixed strains. After eight hours of inoculation period, the required acidity (0.8 to 1.0% lactic acid) is achieved, and the curd is ready for further processing. Preparation of chakka: Chakka/maska is prepared by separating the whey from dahi. Earlier, a 28 inch diameter basket centrifuge at 1,100 rpm was used that produced 80 kg of curd per hour. Now, the process has been upgraded by using a quarg separator. Its capacity is 2,500 kg of curd per hour. Preparation of shrikhand from chakka: Shrikhand is prepared by adding sugar at the rate of 80% of the amount of chakka and mixed in a planetary mixer. Required amount of plastic cream (80% fat) is added along with sugar to chakka so as to give at least 8.5 per cent fat in the finished product on dry matter basis.The processing conditions for the manufacture of shrikhand have been standardized.The final product contains 5 per cent fat, 42 per cent sugar and 60 per cent total solids. The acidity of the product is usually between 1.10 and 1.40 per cent (expressed as lactic acid). The protein content of the product is 10.5 percent.

Upgraded process for shrikhand manufacture.

Starter cultures: The use of right type of culture is essential for the manufacture of shrikhand. A mixed culture containing *Lactococcus lactis subsp. Lactis,Lactococcus lactis subsp. diacetilactis/ Leuconostoc, Lactococcus lactis subsp. cremoris* in the ratio of 1:1:1 may be used. Other recommended cultures are yogurt containing *Streptococcus thermophilus* and *Lactobacillus delbrueckii subsp. bulgaricus* (1:1) or LF-40, a culture compromise of *Lactococcus lactis subsp.Lactis* and*Lactococcus lactis subsp. diacetilactis. Separation of whey from curd:* it is carried out in a quarg separator that can handle 2,500 kg of curd input per hour. This permits to scale up the production to 8-tonnes/ day with installation of other equipment of matching capacity. This process has been further refined for enhancing shelf life of shrikhand by pasteurization. *Mixing of chakka with cream and sugar:* The mixing capacity in the earlier process was limited since the planetary mixer could only handle up to 40 kg/batch.

RELEVANT INFORMATION

Shrikhand is a semi soft, sweetish sour, whole milk product, prepared from lactic fermented curd. The curd (dahi) is partially strained through a cloth to remove the whey and thus produce a solid mass called Chakka (the basic ingredient for Shrikhand). This chakka is mixed with required amount of sugar etc. to yield Shrikhand. It is used as a delicacy in western part of the country, Maharashtra and Gujarat. It shows characteristics flavour and taste due to addition of several ingredients such as cardamom, charoli, nutmug power. Shrikahndwadi is preserved product obtained by the dehydration of shrikhand. The keeping quality of shrikhand can be increased by thermisation treatment.

Precautions:

Use clean muslin cloth. Avoid unhygienic method of handling dahi, Chakka and Shrikhand.

Material required:

1. Dahi
2. Muslin cloth
3. Sugar (powdered form)
4. Nutmug (powdered form)
5. Cardamom (powdered form)
6. Saffron colour.

Apparatus:

1. Kneading machine (Shrikhand Patra)
2. Stand for hanging dahi.
3. Bowl for collection of whey.
4. Stainless steel utensils – 2.
5. Big size spoons.
6. Plastic coated cups.

Procedure:

1. Collect milk acidic dahi (0.7 to 1 percent acidity)
2. Break the curd and place in a muslin cloth (bag)
3. Hang dahi containing muslin cloth on a stand for removal of why for 6 to 8 hrs.
4. The solid mass thus obtained is called chakka.
5. The chakka so obtained is the Shrikhand base.
6. Admix chakka so obtained with sugar @ 35 to 45 % by weight of chakka) by using powdered sugar.
7. Knead well for uniform mixing using a kneading machine (Shrikhand patra)
8. Add colour and flavour (nutmug powder about ¼ small spoon per kg of product and mix it thoroughly.
9. The product so obtained is known as shrikhand. It should be served in plastic coated cups.

Calculations:

Calculate the percent recovery of the product.

$$\text{Percent recovery} = \frac{\text{Wt. Of dahi} - \text{Final wt. If shrikhand}}{\text{Weight of dahi}} \times 100$$

Judging quality of the product:

1. Colour – Texture
2. Body (soft, greasy, lumpy, soggy, buttery)
3. Flavour (pleasant, acidic, putrified off type, aromatic)
4. Taste – (sweetness balancing sweet, sour, too sweet)

PREPARATION OF CHANNA

RELEVANT INFORMATION:

Channa refers to the milk solids obtained by acid hot milk and subsequent drainage of whey. The acids commonly used are lactic or citric in both natural and chemical form. Channa is largely used as a base for the preparation of variety of Bengali sweets such as Rasogolla, sandesh, Pantoa, Channa. It is a rich source of fat and proteins and some minerals. It is recommended for diabetic patient.

Precautions:

1. Use cow milk to obtain good quality of channa.
2. Do not add coagulant when milk is boiling.

Material required:

Milk

Citric acid @ 1 ml of 20 per cent solution of citric acid/lactic acid in one litre of milk piece of clean cloth.

Apparatus:

Still karahi / steel degchi

Chulah

Khunti.

A ladle of Aluminum / steel for removing hot milk from karahi.

Procedure:

1. Bring all the milk to boil by heating directly in the steel karahi over open fire.
2. All the while stir with Khunti and keep at simmering hot (80-82 0 C).
3. Ladle out 0.5 to 1.0 kg of milk into a separate coagulation vessel.
4. Place the coagulant in the coagulation vessel and add the required quantity of milk.
5. Allow the mixture to coagulate completely.
6. Pour the contents in a of clean muslin cloth stretched over another vessel (for receiving the whey).
7. Remove the cloth containing coagulated solids.
8. Tie the solids gently hanging the contents so that there is complete drainage of whey hang up the solids to drain out the whey completely and also to cool the chhanna vat.

Calculations:

$$\text{Percent recovery of Channa} = \frac{\text{Weight of milk used} - \text{Wt. of channa obtained}}{\text{Weight of milk used}}$$

Judging of Chhanna:

1. Colour (Light yellow, whitish).
2. Body (Hard, soft, tough, rubbery sogginess).
3. Texture (coarse, smooth).
4. Flavour (Cooked, acidic, flat, and smokey)

PREPARATION OF RABRI

RELEVANT INFORMATION:

It is a conventional concentrated, sweetened milk product, which has got high appreciation in northern and eastern regions of India. The manufacture of this product is mostly on the cottage scale industry and the method of preparation is primitive and no care is taken as regards the hygienic quality of the product. It is largely prepared by Halawai. The specialty of the product is that it contains several layers of clotted cream.

Precautions:

One should be careful about the danger of open fire heating while concentrating the product.

Material required:

Buffalo / Cow whole milk.

Ground sugar

Apparatus:

1. Large size evaporating pan (karahi).
2. Flat edged scrapper (khunti).
3. Bamboo / cane splinter.
4. Fire chulah.
5. Pan balance.
6. Plastic coated cups.
7. Hand fan.

Procedure:

1. Take 3-4 kg of milk in a karahi and heat it on open fire to 90-100 0 C and maintain it by controlling heating.
2. Do not stir, fan the milk gently by hand fan during heating to help the process of skin formation.
3. Break the skin continuously with the help of splinter and move it to the sides of karahi or collect in the separate pan/breaker.
4. Continue slow evaporation of milk till it reaches to about 1/5 the of its original volume.
5. Add good quality ground sugar 5-6 % by weight of the original milk and dissolve it.
6. Immerse the layers of skin collected on the sides of karahi. At this stage the product is non-homogenous, flakes floating in milk sugar syrup.
7. Heat the whole mass till you obtain homogenous honey comb, textured mass of finished product. Cool it.

PREPARATION OF KHOA

RELEVANT INFORMATION:

In India the quantity of milk available in flush season for exceeds to actual requirements. Under the tropical climate of our country and with limited facilities of storage, the excess of milk cannot be stored or transported as such for marketing. Hence it is imperative to convert into khoa by the producers as a method of utilising surplus milk. Khoa is an ancient indigenous milk product which constitutes an important base for the preparation of various Indian sweet meats.

The surplus milk produced during flush season in converted into khoa by the producers as a method of utilising surplus milk.

There are 3 types (varieties) of khoa viz. Pindi, Dhap, and Danedar, Pindi is used for the preparation of Burfi and Peda. Dhap for Gulabjamun and Danedar for Kalakand. A continuous khoa making machine is having steam jacketed drum heater with a rotary scarper. A milk outlet valve has been designed for large scale production of khoa.

Precaution:

Take at most care while stirring hot milk during desiccation to avoid splashing and excessive charring.

Material required:

1. Buffalo/Cow milk
2. Vegetable parchment paper

Apparatus:

Iron karahi (10 kg capacity) with a top diameter of 40 cm with loop handles on opposite side.

Khunti (65 cm length and 7.5 cm flattened end)

Charcoal chulah (32 cm diameter)

Weighing pan balance

Thermometer

Procedure:

1. Take 2 to 3 liters of milk in Iron karahi.
2. Keep the karahi on non smoky fire or charcoal chulah.
3. Bring the milk to boiling temperature.
4. Start the stirring of milk continuously. Scrap all the parts of the pan covering the milk. Boil the milk vigorously.
5. At this point stir the milk at about 100 rpm. Which would help for constant evaporation of moisture and progressive thickening of milk.
6. The thickened mass shows spurting, abrupt change in colour and consistency at this stage also vigorous stirring and desiccation are continued till the vicous product reaches a pasty consistency and leaves the sides of pan.
7. At this stage give close attention and reduce the fire so as to lower down the temperature to 80-88 0 C.
8. The final product is ready when it shows signs of sticking together.
9. Remove the pan from the fire; work the contents up and down and spreading to the cooler part of karahi. After some time remove the khoa on parchment paper and make circular pat. Weigh the quantity of khoa

PREPARATION OF BASUNDI

RELEVANT INFORMATION

Milk based sweet meats such as Rabri, kheer, khurchan etc. are popular in Uttarpradesh, Punjab, Rajashthan, Bengal and Maharashtra to some extent. Basundi is another sweet meat largely produced in Maharashtra and Gujrat. It is an indigenous product analogue to condensed milk. It is commonly prepared in hotels and on festival occasions in households.

Material required:

Whole milk (preferably buffalo milk) 2 kg.

Powdered cane sugar 300-350 gms.

Pistanchie 10-15 gm.

Cardamom ———————— 1 gm

Borneol ———————————— 50 mg (Edible camphor)

Saffron ———————————— 50 mg.

Apparatus:

1. Wide round bottom pan (preferably khoa pan); Khunti
2. Shegadi or chulah; Double seamed tin containers
3. Tin plate can sealing machine.

Procedure:

1. Take specified quantity of milk in a wide pan.
2. Boil it over a gentle fire.
3. Do not stir the milk too vigorously.
4. Allow to boil slowly to develop a thick skin on the surface.
5. Remove the developed skin periodically by means of a Khunti and set on the cooler sides at the top of the pan. Continue to remove the skin formed due to progressive boiling and collect it on the top sides of the pan.
6. Continue heating of milk to concentrate it to the consistency of the condensed milk (35 % concentrate)
7. Add powdered sugar into the milk and stir it to dissolve
8. At this stage scrap the collected semi dried skins from top of the pan mix into above concentrated milk.
9. Remove the pan from fire, allow it to cool and add flavouring material.
10. Mix cardamom powder as such. Dissolve saffron and borneol in small quantity of water and then add it into contents. Pack the product in double seamed tin containers and cover it.

FEASIBILITY STUDY FOR MILK MARKETING AND PROCESSING

Before one decides to invest in the business of milk marketing and/or processing one should carry out a feasibility study to establish the economic viability of the planned business. this should include a realistic business plan.

THE ESSENTIAL ELEMENTS OF A FEASIBILITY STUDY

The essential elements of a feasibility study should include:

- Establish the amount of milk produced, both in the morning and evening, at the proposed site, throughout the year
- Identify the current market outlets available for milk products in the area
- Determine the average fresh milk and various milk products prices being charged by local producers.
- Test various product samples for taste to determine acceptable products being produced in the proposed area
- Locate sources of energy (fuel-wood, charcoal, electricity, etc. and water)
- Determine the capital investment required (be sure to include land, building, equipment and power).
- Draw up a clear business plan that will establish the viability of the proposed milk marketing or processing enterprise.

Guidelines for business plan preparation

In planning your business, you must examine four major areas:

Description of the business:

a. What type of business are you planning?
b. What product will you sell?
c. What type of opportunity is it (new, expansion, seasonal, year round?)
d. What are the opportunities for growth?

Marketing plan:

a. Who are your potential customers/
b. How will you attract and hold your share of the market?
c. Who are your competitors? How are their businesses prospering?
d. How will you promote your sales/
e. Who will be your suppliers?
f. Where will the business be located?

Organisation plan:

a. Who will manage the business?
b. What qualifications will you look for in a manager?
c. How many employees will you need/
d. How will you manage finances?
e. How will you keep records?
f. What legal form of ownership will you choose and why?
g. What licences and permits will you need?
h. What regulations will affect your business?

Financial plan:

a. What is your estimated business income for the first year?
b. What will it cost you to open the business?
c. What will be your monthly cash flow during the first year?
d. What sales volume will you need in order to make a profit during the first year?
e. What will be your break even level of production?
f. What will be the capital value of your equipment/
g. What will be the total financial needs?
h. What will be the potential funding sources?
i. How will you secure loans?

MILK AND MILK PRODUCTS MARKETING

Until recently the dairy industry in Kenya was characterised by one major processor, the Kenya Co-operative Creameries (KCC). Before liberalisation of the dairy industry in 1992, KCC enjoyed a near monopoly of the Kenya Dairy market. With the emergence of numerous small scale to medium scale dairy processors, the market has become more competitive. In this competitive market, KCC is facing a challenge from the new manufacturers while the new manufactures have to compete first against a well established and large processor like KCC and secondly between themselves for a slice of the same market: the Kenya consumer.

The survival of individual dairy processors will very much depend on how successfully they can win consumer confidence in their products. this calls for knowledge and skills in marketing techniques Most of the newly established small to medium scale dairy processors have very little or no experience in milk marketing.

In order to assist them improve their marketing skills, the training programme for small scale dairy sector under project GOK/FAO/TCP/KEN/ 6611, has prepared this guide on Milk and Milk products marketing to be used for training and by the private small scale dairy processors. The emphasis is on practical approaches to marketing taking into account the real situation existing in Kenya at the present moment.

MARKETING

What is marketing?

Marketing may be defined as "the performance of all business activities involved in the flow of goods and services from the producer to the consumer".

This implies that there are several categories of key players in the marketing chain each with its own vested interests. Consumers want to get what they need at the lowest price possible. producers on the other hand are interested in getting the highest possible return for their milk. Between them, there are market intermediaries or middlemen who perform various marketing functions such as transportation or retailing. Their interest is to make the highest profit possible from their particular business operation.

The Milk Marketing Channel

A study of the milk marketing system in Kenya has shown that there are at least 8 different marketing channels as shown below:

Milk Marketing Channels	Number of intermediaries
Producer-consumer	0
Producer-milk hawker-consumer	1
Producer-processor-consumer	1
Producer-processor- retailer-consumer	2
Producer-dairy co-operative -processor- retailer consumer	3
Producer-milk transporter-processor - retailer-consumer	3
Producer-milk trader-processor-retailer-consumer	3
Producer-dairy coop - milk transporter-processor-retailer-consumer	4

The number of intermediaries involved will have a bearing on both producer and consumer milk prices. The shorter the channel the more likely that the consumer prices will be low and the producer will get a higher return. A notable omission in the milk marketing channel obtaining in Kenya is the absence of wholesalers. Retailers obtain their dairy products directly from processors.

From the consumer point of view, the shorter the marketing chain, the more likely is the retail price going to be low and affordable. This explains why, following the liberalisation of the dairy industry, direct sales of raw milk from producers to consumers (channel 1) or through hawkers (channel 2) has been on the increase despite the public health risks associated with the consumption of untreated milk and milk products. Milk producers may not necessarily benefit from a short marketing chain i.e. milk processors in channels 5 - 6 may be paying farmers the same price as hawkers. However, farmers sometimes prefer selling milk to hawkers because other factors such as prompt payments and inaccessibility to formal market outlets such as producer co-operatives or lack of near by milk processing factory. The biggest disadvantage of direct milk sales to consumers by hawkers is the total lack of quality control and the frequent rate of adulteration of milk with (dirty) water, which is illegal. An efficient milk marketing chain is one which enable farmers to receive at least 50% of the retail price of milk.

MARKETING AND PRICING OF MILK AND MILK PRODUCTS

The price of a product in the market is an important factor influencing consumer demand. Hence to be marketable, a dairy product must be competitively priced. This implies that the costs involved in raw material procurement, processing, packaging, storage, marketing and distribution must be kept as low as possible. generally the price of a dairy product will involve the following costs:

a. Cost of raw milk
b. Cost of raw milk collection and transportation
c. Cost of processing
d. Cost of packaging
e. Cost of marketing and distribution
f. Taxes and tariffs
g. Profit margins at each stage of the marketing channel (Collection, Processing and marketing margins)

In order to arrive at a realistic costing of a product, all those elements involved at each stage must be carefully calculated on a unit basis. This is known as Cost Accounting.

Market function	Cost element
1. Raw milk procurement	Cost of raw milk; labour; materials etc.; collection margin
2. Transportation	Transport cost; labour; materials and equipment; transport margin
3. Processing	Raw materials; machinery and equipment; labour; packaging; energy; taxes; marketing and distribution; processing margin
4. Marketing and distribution	Transport; labour; materials; rent; retail margin

The cost can be broadly categorised as fixed costs and variable costs. Fixed costs include things like depreciation of equipment and buildings while variable cost include direct expenses such as raw material; marketing expenses; overhead costs.

It is important that all the cost elements are included in the calculation of the market value of the product. Overpricing can lead to uncompetitiveness of the product while under pricing can cause financial loss and eventual collapse of the business.

MARKETING INFORMATION SYSTEM AND RESEARCH

Information is required at all levels in the marketing channel. Before you decided to process and market any dairy product, it is important to know the potential market for each particular product. This is important to enable the processor to know which types and when, where and how much of each product to manufacture and market. It is very crucial because unless goods can be supplied in the right form, place and times, consumers may not be able to buy. This then requires securing and utilising market information.

Marketing information should address the following:

- Area to be covered
- Price information (Price variations, price for premium quality discount price etc.)
- Number and type of consumers (market segmentation)
- Current and future product supply levels
- Type and number of competitors

In the absence of comprehensive marketing information system such as is the case in many developing countries it may be necessary for each individual processor or through their organisation to organise the gathering and dissemination of such information. Short market survey and/or Consumer studies are useful tool for gathering such information.

PRESERVATION AND PROCESSING AT HOUSEHOLD LEVEL IN MILK PRODUCTS

Milk is a valuable nutritious food that has a short shelf-life and requires careful handling. Milk is highly perishable because it is an excellent medium for the growth of microorganisms – particularly bacterial pathogens – that can cause spoilage and diseases in consumers. Milk processing allows the preservation of milk for days, weeks or months and helps to reduce food-borne illness. The usable life of milk can be extended for several days through techniques such as cooling (which is the factor most likely to influence the quality of raw milk) or fermentation. Pasteurization is a heat treatment process that extends the usable life of milk and reduces the numbers of possible pathogenic microorganisms to levels at which they do not represent a significant health hazard. Milk can be processed further to convert it into high-value, concentrated and easily transportable dairy products with long shelf-lives, such as butter, cheese and ghee.

Processing of dairy products gives small-scale dairy producers higher cash incomes than selling raw milk and offers better opportunities to reach regional and urban markets. Milk processing can also help to deal with seasonal fluctuations in milk supply. The transformation of raw milk into processed milk and products can benefit entire communities by generating off-farm jobs in milk collection, transportation, processing and marketing.

Milk producers in the tropics apply their ingenuity to cope with climatic and logistical adversities. Being a subsidiary occupation, the average milk producer produces in small quantities. The market is not always within easy reach for delivering fresh raw milk and local processors, if any exist, are constrained to take milk in excess of processing ability as they may be plagued by problems like power failure or transport breakdown. Producers, therefore, have adopted, over the generations, a variety of methods to minimise the loss in value of fat and SNF (solids-not-fat) in liquid milk. The products developed, either for direct consumption or as a base material may have some local variations, in terms of additives, but follow certain fundamental rules of preservation and conservation through, heat desiccation, fermentation, coagulation and clarification. These rules are followed in an effort to recover the total solids to the maximum extent, but take place through processes and technologies which shock modern standards but remain adequate and appropriate to local situations, resources and food-habits. The processors, similarly, adopt small-scale processing technology which is adequate to process limited marketable surpluses of an area, often in the form of a chiller or a bulk cooler, as independent units to cater to the local demand for liquid milk or as feeders to bigger units of the area. As for small-scale processing, it is believed that 'it is in fact possible to manufacture the highest quality dairy foods with extremely simple tools and means. It is also true that most dairy foods can be made with the help of relatively few basic processes provided that the manufacturer has the necessary skills and experiences. Complicated machinery and sophisticated processing methods have not been invented to make better dairy products, but to save labour and nothing else'.

In most of the tropical countries, a fraction of the total milk production enters the formal processing channel for reasons. In India, for example, even after nearly three decades of organised milk procurement and market intervention under the Operation Flood programme, of the estimated 30 per cent of the total production which is marketed by the village producers (after meeting the local consumption needs, of both the producers and the non-producers), only 35 per cent is collected by the organised processing sector and the rest by the traders of all hues.

According to another estimate, 28 per cent of total production is converted into *ghee* and another 20 per cent into milk products, such as *dahi* (curd), *khoa* (dehydrated milk) and a variety of milk-sweets, to enhance shelf-life. The villagers try to conserve milk in its freshness, before it becomes sour which makes possible better recovery of milk solids in the products that are made. Much Indian milk production is seasonal in character as nearly half the production is by buffalo which are seasonal calvers. Seasonal surpluses are converted into the above products for which there is a premium in India. The pattern is much the same in Pakistan, Bangladesh, Nepal and Sri Lanka. A generalised flow of milk from the producer to the consumers and the prevailing processes of conservation in the Indian sub-continent.

However, Operation Flood provided the Indian disorganised milk producers (the majority being landless or small/marginal holders) in distant and remote villages, with a ready market for their produce year-round and at remunerative prices. This encouraged them to produce and sell more liquid milk. Village-level primary co-operatives collect milk from their members and transported it to the District Union's modern dairy plant, unless local circumstances demand intermediate chilling or cooling facilities, for processing into a variety of milk and milk products.

Introduction of refrigerated road and rail milk tankers make collection and haulage of liquid milk over long distances easier. The impact of providing access to the market is evident from the steady increase in milk procurement under Operation Flood. In 1970-71 it was 0.5 million litres a day and increased to 12 million litres a day in 1996-97. Expansion in the organised processing and marketing fields had a similar effect on smallholders participation in dairying in Kenya where, in addition to co-operative creameries, rural separating centres, and where suitable, milk cooling and collection centres, were set up for cream production to which local farmers supply their milk even in such small quantities as half a gallon.

In some other countries, consumption demand for sour milk has eased problems of fresh milk marketing. In northern Nigeria most of the milk is locally consumed, either fresh or sour with a relatively small fraction in the form of local cheese. In Kaduna, 70 per cent of the town's surveyed households consumed only *nono* (sour milk) not fresh milk, because it sours rapidly in hot weather. *Nono*, the most preferred milk-type product, is made from skimmed milk and usually sold with *fura* (doughy patties of sorghum or millet mixed with spices and sugar). *Lamudam* or *kindirmo* (yoghurt, more often found in remote rural areas) is thicker than *nono* and made from whole milk. *Cucu* or *wara* is a soft, wet, feta-type cheese made from whole milk. This is sold in small cones, sometimes fresh, but more often fried. Markets also exist for*Nebbam* or *Man shanu* which is a local butter.

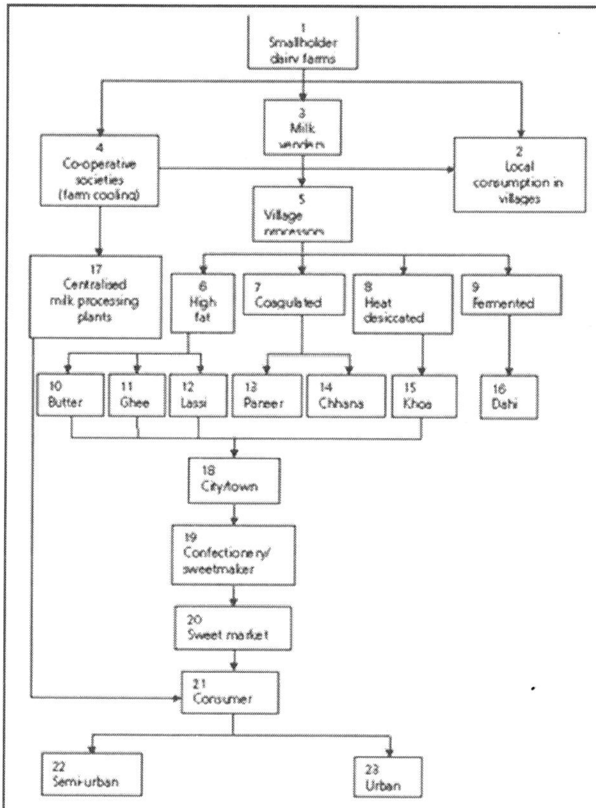

Milk movement scenario in the Indian sub-continent.

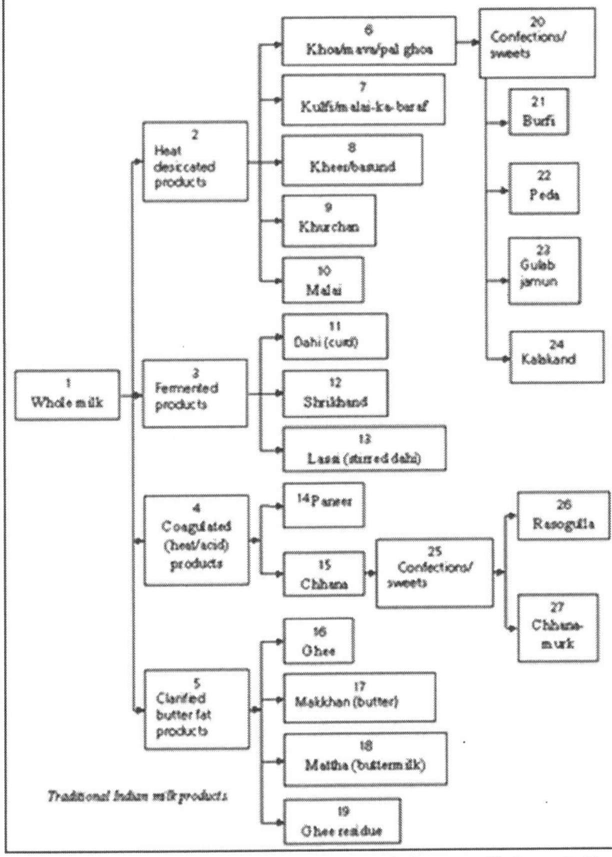

Traditional Indian milk products

The sour milk forms an important part of the traditional diet, fresh milk is seldom drunk alone. Often naturally soured, the milk coagulates to produce a thick curd called *amasi*. This is often eaten with *Sadza*, a thick maize porridge. In semi-arid Borana plateau of Ethiopia, milk is typically stored to produce *ititu* or *irgo* (in Amharic) which is a generic term for either (1) short-term (five day) fermented milk for consumption or making butter (2) longer-term (up to 90 days) fermented milk particularly kept for guests. They also make ghee which can be stored for up to three years. The household processing system can process from a low of about two litres at a time to 100 to 500 litres of fresh milk, in batches of up to 15 litres.

A key to dairy profitability is the extent of fat recovery. A review of milk separation by traditional hand churning, a wooden paddle agitator and a centrifugal separator suggests that the latter gave the best fat recovery but is too costly for small farmers and would be profitable for production of more than 12 litres a day all year round. This would be appropriate for some co-operatives and larger farmers. In Ethiopia's informal system, a smallholder sells surplus supplies to neighbours or in the local market, either as liquid milk or in the form of butter or a cottage-type cheese called *ayib*. In the highlands of Ethiopia, milk produced by smallholders is used for family consumption and the production of butter and *ayib*. For butter-making, milk is collected over a period of three or four days in a clay pot. When the milk has soured and sufficient milk has been collected, the clay pot is shaken back and forth, for about two to three hours, depending upon such factors as temperature, quantity and fat content and the acidity of milk. This continues until butter granules are formed.

An innovation developed by the International Livestock Research Institute, formerly the International Livestock Centre for Africa, modified a wooden internal agitator that can be fitted to the usual clay pot, to reduce churning time from an average of 139 minutes to an average of 57 minutes (59 churnings) while reducing the average fat content of the buttermilk from 1.1 per cent to 0.36 per cent.

In these processes, residual buttermilk is heated, precipitating the casein and some of the remaining fat, and the coagulated fat and protein is separated from the whey to produce *ayib*. The price of *ayib* is about one-seventh that of butter. The monetary advantage of extracting the maximum amount of fat from milk and converting it into butter is apparent. In Bamako, Mali, surveyed urban households were reported to be consuming Sirime (ghee) and sour milk, sweetened or otherwise.

In Debre Zeit, Ethiopia, because of availability of fresh milk in the majority of the rural households, only two per cent of smallholders sold fresh milk. Butter, used in rural areas for cosmetic and cooking purposes, is sold by almost 67 per cent of smallholders. About 64 per cent of the smallholders consumed all the cheese they produced and the remaining 36 per cent consumed and sold the produced cheese to different extents. Ghee has a considerably long shelf-life than butter but customers prefer to buy butter and make their own ghee. Spices may also be added to butter to improve its shelf-life but this method is usually confined to butter for use by smallholders.

Even when slightly rancid, butter still has a market value. Urban consumers of highland origin use butter for making traditional Wots that are eaten with bread-like Injera made from Teff (*Eragrostis tef*). A degree of rancidity is actually desired to improve the flavour of *wot*. In Madagascar, depending on the region (southern Madagascar has a tradition of curdled milk consumption) and on consumer demand, producers deliver either fresh or curdled milk. When direct sales to consumers are not possible, producers resort to middlemen which include cycling milkmen, small-scale butter and cheese producers, collection centres and dairy plants. In northern Tanzania, 57 per cent of the milk protein consumed was in the form of fermented milk . In the rural areas of Uganda, a certain amount of ghee production was reported, mainly for household consumption or local trading.

SMALL-SCALE PROCESSING UNITS

Milk must reach a cooling facility or processing plant within two to five hours of milking, or chemical changes occur that make it unsuitable for processing. In arid and semi-arid regions of Africa and the Indian sub-continent, the amount of milk available from farms within two to five hours of any point is not likely to be sufficient for any but small plants. In some parts of Kenya, the amount of milk that can reach commercial factories falls by as much as 20-30 per cent during the wet season because of difficulties with roads . In Palestine, it was observed that farmers who comprise 40 per cent of the population are often left with milk production which they are unable to send for processing because of the absence of a transport and processing infrastructure. In India, co-operative milk plants in the low-lying Gangetic plains are unable to procure milk from many village societies in the rainy season as floods cut off roads. In southern Chile, it was observed that, some dairy farms could not deliver milk in winter due to poor road conditions which affected the operations of the processing plants, with volume of reception showing seasonal fluctuations in a ratio as high as 1:5.

Thus, small scale processing becomes appropriate in situations where transportation is a constraint, primary production is dispersed, quantity of marketed surplus is limited or immediate neighbourhood demand is low. Since, such plants process the milk in the production area itself, they have lower transportation/handling costs than big collectors and provide better returns to the producer by value additions. An added advantage is that when the power fails, alternative methods employing manual labour are used to save the milk and thus reduce the risk of wastage.

Requirements of lower levels of energy, skill, maintenance and repair also have a positive impact on local milk production. Cooling centres are reported to have a positive impact not only in providing access to marketing but also making possible more frequent milk collection. A small-scale plant near a small town in Northern India was reported to be processing 1500 kg of milk a day, using simple methods that are mostly hand operated and produce pasteurised milk, sweetened condensed milk, fresh butter, ghee, cheeses and cream caramels .

In Peru, setting up of rural cheese plants in distant villages helped improve milk quality owing to shortened distance between producers and the plant. In Peru and Ecuador, 60 rural cheese factories have been established at a cost of US$ 2000 in locally made equipment, US$ 1000 for furbishing buildings and US$ 100 as working capital. Simple cheese technology is available in manuals. The Netherlands has supported cheese making in parastatal farms in Tanzania and Sri Lanka and at the co-operative level in Colombia. In Ecuador, Cheese Co-operatives were formed in areas not reached by middlemen for milk collection and where farmers could collect, within a three km radius, at least a daily quantity of 300 litres of milk. In Afghanistan, small processing plants collecting milk within a radius of maximum of two hours of transportation time have helped collection of good quality milk without cooling. It proved to be easier to build a second processing centre than to centralise milk processing with the help of expensive and unreliable chilling centres .

Alternatively, as in East Africa, a series of cooling facilities collect producers' milk and despatch it to a larger processing plant. But such a system comes under stress when the cooling facilities or transport to the plant fails and when the large plant itself has problems with machinery, power or inputs such as packing materials .

In Africa, small-scale processing could be extended to traditional products such as sour milk, ghee and white cheese in brine, for which the local population has the skills and for which there are local markets. The cheese industry in Sudan's White Nile Province specialises in making soft white cheese using methods that were introduced at the beginning of this century. Usually one factory serves one village, handling a daily average of 275 litres of fresh milk, operated by unskilled workers except for the head worker. There were two factories in 1935, increasing to 70 in 1975. Most of their output is sold in Khartoum and other cities . However, small-scale decentralised processing has certain dis-economies of scale. Processing of less than 2000 litres would not be viable, and product quality may differ from plant to plant. In Tanzania, small processors need foreign exchange to obtain inputs such as rennets and starter cultures.

To overcome the problem of low capacity utilisation due to low volumes or seasonal supplies, it is suggested that dairy plants could make use of local fruit juices and vegetable extracts, as they do in Asia. 'Vegetable milks' could supplement animal milk during the dry season when animal milk is in short supply. Vegetable oils could be mixed with milk to produce a high-calorie, low-cost food. Tropical fruit juices and bean extracts could be used to flavour and acidify milk to produce a high-protein, high-vitamin food. As one researcher observed, 'One can explore a great variety of possibilities as soon as one accepts that milk is not the only raw material that can be processed in a dairy'.

TRADITIONAL MILK CONSERVATION AND PROCESSING PRACTICES IN THE INDIAN SUBCONTINENT

Because of the generally prevailing unorganised transport, processing and marketing facilities, the people of the tropical countries have mastered the art of conserving or processing the limited quantity of milk they produce. The household and small-scale processes prevailing in the Indian sub-continent, countries which share common food-habits and preferences.

Indigenous dairy products of the Indian sub-continent could be broadly classified into the following four categories:

1. Heat desiccated/unfermented concentrated whole milk products
2. Fermented indigenous products
3. Coagulated indigenous dairy products, and
4. Clarified butter fat indigenous dairy products.

HEAT-DESICCATED/UNFERMENTED/CONCENTRATED WHOLE MILK PRODUCTS

Khoa (alternatively called *mava, khava* or *palghoa*), is a partially dehydrated whole milk product, prepared in a batch of four to six litres in a traditional open shallow pan over a direct fire, by continuous heating and manual stirring-cum-scraping until it reaches a semi solid (doughy) consistency. Pan-contents, *khoa-pat*, are removed from the fire and worked into a solid mass of different grades of texture (smooth, homogeneous, granular), consistency (thick, sticky) and moisture (31-44 per cent), depending on the type of use it would be put to for making a final product. Three grades of commercial *khoa* are made:

1. *Pindi khoa*: 21-26 per cent fat, 31-33 per cent moisture
2. *Dhap khoa*: 20-23 per cent fat, 37-44 per cent moisture
3. *Danedar khoa*: 20-25 per cent fat, 35-40 per cent moisture.

The shelf-life of *khoa* is two to four days under ambient conditions and three weeks under refrigerated conditions. Traditionally, *khoa-pats* move from village to towns wrapped in leaves and put in bamboo baskets. The shelf-life can be enhanced by using vegetable parchment paper wrappers, plastic (polyethylene) film bags/pouches, laminated (preferably aluminium coated) pouches, tin plates, or cans, in pack sizes varying from 0.5 (or even less) to 1.0 kg. *Khoa* contains fairly large quantities of muscle building proteins, bone-forming minerals and energy-giving fat and lactose. It is also expected to retain most of the fat-soluble vitamins A and D, and also fairly large quantities of water soluble B vitamins contained in the original milk. Additives to sweetmeats further increase the calorific value of the product. Using cereals, wheat flour, sugar, flavours and other additives, depending upon the local preferences, a wide range of milk sweets/confections/delicacies are made from *khoa*, on a small-scale and commercial basis:

- *Gulabjamun*: Balls prepared by kneading *pindi khoa* as base material with additives like refined wheat flour, semolina and baking powder followed by deep frying in cooking medium and further dipping in a sugar syrup.
- *Burfi*: *Khoa* based sweet, prepared by heating *khoa* on medium heat, topped by nuts such as *pistachio, cardamon* or a layer of chocolate.
- *Kalakand*: Prepared from Danedar Khoa, with typical granular texture and added aromatic spices.
- *Peda*: Ball shaped (top and bottom flattened), prepared from *dhap khoa*, topped with nuts.

Technological innovations in Khoa-making

The primitive process of *khoa* making in mild steel shallow open pans has been scaled up by different semi-continuous and continuous machines which are used in places where the quantity of milk is sufficiently large for*khoa* making. A semi-continuous *khoa* making machine was developed by Banerjee et al. in 1976. The plant consisted of a scraped surface heat exchanger and two open semi-jacketed pans with reciprocating spring loaded scrapers. The scraper adjusting mechanism was developed by Sawhney et al in 1980 providing a semi-jacket, shallow open pan and using a swinging hanger type scraper for stirring during the desiccation process.

The Conthern-Convap System developed by Alfa-Laval has also been used for manufacture of *khoa*. A conical vat machine has also recently been developed. The latest innovation is the development of an inclined scraped surface heat exchanger for continuous *khoa*-making by National Dairy Development Board . The plant comprises a balance tank, a positive displacement pump and an Inclined Scraped Surface Heat Exchanger (ISSHE). Milk concentrate is used as feed. The sensory characteristic of *khoa* prepared by ISSHE is similar to those of the traditional product and the chemical composition and rheological properties of *khoa* maintain uniformity during continuous operation. ISSHE is compact and simple and is being used on a regular basis.

TECHNOLOGICAL INNOVATIONS IN GULABJAMUN MANUFACTURE

The continuous process line developed by the NDDB, and being used by co-operative dairies in India, for large-scale manufacture, comprises of dough formation, portioning, shaping, frying, sugar syrup soaking, cup filling and sealing.

Kheer or *Basundi* is a cereal-based heat desiccated traditional dairy product, prepared by partial dehydration of whole milk in a traditional open shallow pan over a direct fire, together with sugar and usually rice or occasionally semolina. Consumed directly as a dessert, it contains all the solids of milk in an approximate twofold concentration plus additional sugar, with high food and nutritive value. Additives increase the calorific value of the product (its biological value is 77 and digestibility is 97 per cent). On an average it contains 67.02 per cent moisture, 7.83 per cent fat, 6.34 per cent protein, 8.45 per cent lactose, 1.41 per cent ash and 8.95 per cent sugar. Its average shelf-life, two to three days at 37_C and 10 to 15 days at 4_C, can be increased significantly by addition of nicin to the hot product, at the end of manufacturing process and before packaging.

Rabri is a concentrated, sweetened whole milk product, containing several layers of clotted cream. While the milk is slowly evaporated, without being stirred, at simmering temperature in a traditional open shallow pan over an open fire, pieces of skin which form on the surface of the milk are continuously broken up and moved to the cooler parts of the pan. Sugar is added when the volume of milk has been considerably reduced, layers of clotted cream are immersed in the mixture and the finished product is obtained by heating the whole mass for a short period.

It contains all the milk solids in an approximately five-fold concentration, with additional sugar. Consumed directly, it has a high food and nutritive value, with approximately 20 per cent fat, 10 per cent protein, 17 per cent lactose, 3 per cent ash, 20 per cent sugar and 30 per cent moisture.

Malai is usually made by simmering large quantities of milk (about 10 kg) until a thick layer of milk fat and coagulated protein forms on the surface. This is skimmed off with a flat ladle and laid aside to cool. The process is repeated twice, when most of the fat has been removed. The product is smooth and white in appearance, tastes like clotted cream and makes up about 20-25 per cent of milk. It is sold and consumed without any addition of sugar.

Kulfi is a heat desiccated frozen milk product (indigenous ice cream). While the milk is boiling, sugar is added, the product is concentrated at a ratio of approximately 2:1. Malai, crushed nuts and flavouring (commonly rose or vanilla) is added to the cooled concentrate, the mix is placed in moulds (triangular, conical or cylindrical) of various capacities made of galvanised iron sheets. The moulds are closed on top by placing a small disc over them, the edges made air-tight with wheat dough (modern moulds are made of plastic, generally conical in shape with screw-cap plastic tops) and the moulds are frozen in a large earthen vessel containing a mixture of ice and salt (1:1). A similar product is malai-ka-baraf, prepared by freezing a mixture of malai, sugar, nuts and essence. The product is plastic in nature, with a crunchy texture and delicate nutty, caramel flavour.

Khurchan is a concentrated, sweetened whole milk product prepared in a traditional open pan by simmering the milk without stirring so as to allow for simultaneous formation of a thick creamy layer of skin on the surface of the milk and slow evaporation of water. Sugar is added to the concentrate followed by a thorough mixing of the product. Average composition is 23.6 per cent fat, 15.4 per cent protein, 14.9 per cent lactose, 15.2 per cent sugar, 3.0 per cent ash and 27.9 per cent moisture. It contains all the milk solids in an approximately five-fold concentration, together with an addition of sugar, so its food and nutritive value is very high.

FERMENTED PRODUCTS

Indian households partially extend the life of milk from 12 to 24 hours by simple heat treatment which destroys most microflora. Temperatures of 20-40_C in the warm parts of the year are most favourable for the growth of micro-organisms which bring about spoilage of milk within a few hours of milking. So the simplest way of preserving milk for human consumption in a tropical country is to allow it to sour which checks putrefactive changes while giving the milk an acid taste which is particularly refreshing in a hot climate. A variety of fermented milk products are made and consumed in the Indian sub-continent.

Dahi (curd) is obtained from pasteurised or boiled milk by souring, natural or otherwise by a harmless lactic acid or other bacterial culture, cane sugar may be added if preferred. Households use the previous day's *dahi* as the starter for their preparation. Today bacterial cultures in the form of ampoules are available in local markets for *dahi* preparation. If it is good quality, it does not have any wheying off and has a clean flavour. Confectioners (Halwais) concentrate the milk before inoculation and set the *dahi* usually in circular earthenware (modern packaging includes glass bottle and plastic or plastic coated cup). It has the same percentage of fat and SNF as the milk from which it is prepared. Average composition from whole milk is 5-8 per cent fat, 3.2-3.4 per cent protein, 4.6-5.2 per cent lactose, 0.70-0.72 per cent ash, 0.11-0.5 per cent lactic acid 85-88 per cent water. It is used for direct consumption either as such with salt or sugar or in the form of a popular beverage called *lassi*. *Lassi* is a whitish, viscous fluid with creamy mouth feel and rich aroma of diacetyl, it is mildly to highly acidic a property obtained on stirring Dahi with about 10-20 per cent water and sugar or salt, it is chilled when consumed in summer and directly consumed. Those who do not like drinking milk would consume Dahi readily; its biological value is 66 and digestibility 98 per cent. It contains antimicrobial and anti-carcinogenic factors, stimulates the immunological system and helps reduce blood serum cholesterol, with possible therapeutic value in stomach and intestinal disorders, due to its antibiotic content. The recommended storage temperature is around 5 to 10_C because it has short keeping quality at room temperature. On prolonged storage, it becomes highly acidic, accompanied by whey formation, making it unfit for human consumption.

Misthi doi is a sweetened, fermented milk product, traditionally prepared from cows' milk and consumed directly. It is nutritious, palatable and considered to be a delicacy in the eastern part of the country. It contains all the milk ingredients in a concentrated form together with sugar and caramel. Milk is mixed with sugar, concentrated by boiling, cooled and inoculated with culture (usually the previous day's dahi is used), and set in earthen pots. Good quality is characterised by firmly set curd with a fat layer at the top, slightly brownish appearance and a sweet taste with a mild and pleasant acidic flavour. Its average composition is 4.0 per cent fat, 13 per cent SNF, 18 per cent sugar, 0.1 per cent caramel and 64.9 per cent moisture. It is packed in earthen wares, poly-styrene cups (modern packaging), and the shelf-life at cold room temperature of 4_C is one week.

TECHNOLOGICAL INNOVATIONS IN dahi MANUFACTURE

Production of *dahi* is now being attempted on an industrial scale in a limited way. In one suggested process milk is standardised and pre-heated to 60_C and homogenised at 176 kg/cm^2. Final heat treatment is given at 80-90_C for 30-50 minutes, cooled to 22-25_C, inoculated with a lactic culture and added suitable cups. The cups are placed in an incubator at 22-25_C for 16 to 18 hours. *dahi*obtained in this way is stored at 5_C. Its keeping quality is about a week at refrigerated temperatures. Attempts have been made to extend its shelf-life by thermisation treatment .

It has been claimed that bubbling of $CO2$ at one psi per minute before fermentation, yields a shelf-life of 15 to 30 days at ambient temperature . Attempts have also been made for the production of dried *dahi*by spray drying . Such processes have been developed for the production of a dried *dahi* which could be set in a firm gel structure upon reconstitution. However, such a product lacked the rich aroma of a naturally fermented product.

Shrikhand is a semi-soft sweetish-sour whole milk product prepared from lactic fermented curd (*dahi*). The curd is partially stained through a cloth to remove the whey and thus produce a solid mass called *chakka*(containing an average 63.2 per cent moisture, 14.7 per cent fat and 0.8 per cent lactic acid). *Chakka* is mixed with the required amount of sugar, flavouring agents like safron, and cardamom, to produce *shrikhand.*

Consumed directly, it could be further desiccated over an open pan to make *shrikhand wadi* sweet. Also consumed directly, this contains an average of 6.5 per cent moisture, 7.4 per cent fat, 7.7 per cent protein, 15.9 per cent lactose, 0.8 per cent ash, 62.9 per cent sugar and one per cent lactic acid (as obtained from laboratory made samples from buffalo milk products). Shrikhand contains the total solids of *dahi* including protein and only a part of whey containing lactose, lactic acid and water soluble vitamins. Added flavours, aromatic spices, dry fruits and nuts, increase the caloric value of the product. It is packed in earthenware cups and in modern packaging like plastic or plastic coated cups. Recommended storage temperature is 5 to 10_C as it has short shelf-life at room temperature.

TECHNOLOGICAL INNOVATIONS IN SHRIKHAND MANUFACTURE

Shrikhand is also being manufactured on an industrial scale, in the process line developed by NDDB. In this process, skim milk curd made with a standard culture under controlled conditions from pasteurised skim milk, is centrifuged in a continuous quarg separator to produce *chakka*. This is then mixed with cream, ground sugar and flavours in a scraped surface heat exchanger for manufacture and pasteurisation of the product. It is then filled in a form fill and seal machine in a semi-aseptic environment before despatch for retail trade. Incorporation of modern technology and radical innovations in the traditional practices has not only made possible the manufacture of a safe, uniform, high quality product but has also enabled the organised sector to manufacture the product on an industrial scale and expand the market segment and improve its financial performance.

Coagulated (heat/acid) products

Milk is coagulated (separated out into solid and liquid portions) under varying conditions. This is undesirable when brought about by uncontrolled agencies. However, the same phenomenon has been exploited to concentrate the whole milk solids for edible purposes. Some products made on the basis of coagulation.

Panir is an acid coagulated or rennet coagulated, small sized soft cheese. Traditionally, milk is boiled in a vat and coagulated with citric acid/malic acid at 70 to 85_C. A coagulum of casein-whey-protein complex with entrapped fat is formed, free whey is drained off from the vat and the coagulated mass is pressed in cloth lined hoops. The coagulum knits together into a compact spongy mass under pressure. After removal from the hoops, blocks are placed in chilled water for firming.

Blocks are then cut into smaller pieces of suitable size for retail sale. Approximate composition is 55-70 per cent moisture, 22-26 per cent fat, 15-20 per cent protein, 2.0-2.5 per cent lactose and 2-2.2 per cent ash. The product may be losely packed in polythene bags but could also be packed parchment paper. Normal shelf-life is one to two days for the traditionally made product or six days in refrigerated storage (5 to 10_C) though the freshness is lost after three days. Frozen *panir* can be stored for up to three months at 20_C and is as good as fresh.

Panir is used for direct consumption and also forms the base for a variety of culinary dishes and stuffing material for various vegetable dishes, snacks and sweetmeats. Though it can be produced from both buffalo and cow milk, *panir* from buffalo milk is preferred. It forms the base for a variety of culinary dishes, stuffing material for various vegetable dishes, snacks and sweetmeats. Different local varieties of Panir are:

Surati panir/Cheese is a rennet coagulated product and the best of the few indigenous varieties of panir or cheese. Cakes of *surati* cheese are carried over long distances by rail and road, floating in lactic whey in large sized earthen containers and sold directly to consumers.

Dacca cheese/panir is a small, medium-pressed cheese. Milk is first coagulated with rennet, then broken by hand and filled into small wicker baskets, whereas whey is pressed out by placing a board with a weight on the baskets. After 10 to 14 days, when the cheese pieces become sufficiently dry and develop a dried surface, the cheese is smoked, generally over cow dung smoke. The smoked cheese keeps well for one to two months. The combined tarry and furaldehyde flavour imparted by smoking is not liked by some people. During storage there is no breakdown of protein or fat, so the keeping quality is good.

Bandal cheese/panir is a small, soft type of Panir made from either cow's milk or low fat cream. It is prepared in a more or less similar manner as *surati panir*, but is normally 'smoked' under crude indigenous conditions.

Chhana constitutes one of the two chief bases (the other being *khoa*) for the preparation of indigenous sweetmeats. It represents milk solids obtained by acid coagulation of boiled hot whole milk and subsequent drainage of whey. The acids commonly used are lactic or citric in both natural and chemical forms. Cow/mixed milk is brought to the boil by heating it directly in a traditional open shallow pan over an open fire, all the while stirring it with a ladle and later keeping it simmering hot in the pan. This hot milk is ladled out in batches of 0.5-1 kg into a separate coagulation vessel, either already containing, or to which is promptly added, the required quantity of the coagulant (normally cleansed sour *chhana* whey maintained in a large earthen vessel). The mixture of milk and whey is stirred with a ladle and when completely coagulated, poured over a piece of clean muslin cloth stretched over another vessel for receiving the whey. The process is repeated until all the milk is used up. The cloth containing the coagulated solids is tied up into a bundle without applying pressure and hung up not only to drain out whey completely but also to cool the contents (*chhana* pat) which are used in sweet preparations. The average chemical composition from cows milk is 53.4 per cent moisture, 24.8 per cent fat, 17.4 per cent protein, 2.1 per cent lactose and 2.1 per cent ash. It has a fairly high fat and protein content, contains some minerals, especially calcium and phosphorus. Its food and nutritive value is fairly high and it is a good source of fat soluble vitamins A and D. For its high protein and low sugar content, *chhana* is highly recommended for diabetic patients. Milk sweets from Chhana are:

Rasogolla: Balls made by kneading *chhana* with semolina and baking powder, cooked in sugar syrup.

Sandesh: Small cubes made by kneading *chhana* and cooking in sugar syrup over medium fire.

Chhana-murki: Small cubes of *chhana*, cooked in boiling sugar syrup till they attain sufficient firm body and close knit texture.

Rosmalai: Chhana balls cooked and served in thickened milk, sweetened with added aromatic spices and nuts.

Clarified butter-fat products

Products made in this method of preservation of milk form an integral part of the diet of the people of the Indian sub-continent.

Makkhan or butter normally obtained by churning whole milk *dahi* (curd) with crude indigenous devices, is essentially an input to the ghee (clarified butter) making process at household level though people also apply it directly on chhapatis (leavened bread). Buffalo milk yields whitish product, whereas cow milk product is yellowish. *Dahi* is mixed with cold water and churned in earthen vessels with a wooden beater, grains of Makkhan appearing on the top are scooped out and patted into a smooth compact mass having 18-20 per cent moisture, 78-81 per cent butter fat, 1.0-1.5 per cent SNF and not more than 0.2 per cent lactic acid. It contains vitamins A, D, E and K (fat soluble vitamins), small amount of essential fatty acids, arachidonic and linoleic acids. Parchment papers are used in packaging, for refrigeration and storage. Keeping quality is low under existing rural conditions but improved methods of production can enhance the keeping quality.

Mattha (butter milk) is the whitish, flavory, highly acidic fluid left over after churning Makkhan for the preparation of ghee in the unorganised, traditional, household sector. It is directly consumed, with or without any salt added.

Ghee, the butter-fat prepared chiefly from cow or buffalo milk, is the most common milk product in the Indian sub-continent. It is used as a cooking or frying medium, and is also consumed directly (with rice or *chhapatis*) apart from being used in confectionery and in traditional medicines. Traditionally, the household ferments whole milk into *dahi* and churns out *makkhan* (butter). Fresh or accumulated over days, *makkhan* is clarified at 105-145_C in a suitable open mud pot or metallic vessel, stirred continuously on a low fire to drive out all the moisture. The vessel is then removed from the fire and, on cooling, the residue settles down, and the clear fat is decanted into a suitable container. Ghee has somewhat longer keeping quality than *makkhan* and can be transported over long distances. Modern dairy plants manufacture ghee on a large scale. The development of an oxidised flavour or tallowiness in ghee is accelerated at higher storage temperatures. At household level, its storage temperature may vary from 5-38_C (usually recommended at 21_C) though refrigeration delays acid development and prolongs the keeping quality by rendering the product greasy and pasty. The average chemical composition of ghee from cow milk is 99-99.5 per cent fat, not more than 0.5 per cent moisture, 3.2-7.4 micro g/g unsaponiable matter in terms of carotene, 19-34 I.U./g Vitamin A and 26-48 micro g/g tocopherol. Free Fatty acid is limited to 2.8 per cent in terms of oleic acid (Agmark standard, that is, Indian Standards for Ghee). In buffalo milk ghee, milk fat, moisture and free fatty acids are the same as in the case of cow milk ghee; unsaponiable matter in terms of Vitamin A is 17 to 38 I.U./g and tocopherol is 18 to 31 micro g/g. Ghee contains Vitamins A, D, E and K, small amounts of essential fatty acids arachidonic and linoleic acids. Because it is susceptible to deterioration due to exposure to light, air and metals, tin-coated (non-toxic and non-tainting) containers are used for packaging. Food-grade plastic containers and polyethylene pouches are also used.

INNOVATION IN PROCESSING EQUIPMENT FOR GHEE MANUFACTURE

In a continuous process developed by the National Dairy Research Institute (NDRI), the equipment consists of receiving-cum-heating vat, gravity separator (which becomes optional when making the product from cream), scraped surface heat exchangers coupled with vapour separators and positive displacement pumps to move the raw material through different units. A continuous ghee making plant has been recently developed by NDDB wherein the concentration of fat and breaking of fat in water emulsion is done mechanically with the help of centrifugal force in the clarifixator and concentrator and use of heat is limited to development of flavour and removal of traces of moisture. The plant developed by NDDB requires little energy and can be adopted for much larger volumes. The process also results in lower losses of SNF and better recovery of fat while reducing thermal pollution of environment.

Ghee residue refers to the charred (burnt) light to dark brown residue which is obtained on the cloth strainer after the ghee is filtered. The average composition of the residue is, depending on whether *makkhan* is derived from cow or buffalo milk, 32.4-33.4 per cent fat, 32.8-36 per cent protein, 12-15.4 per cent lactose, 5.2 per cent ash and 13.4-14.4 per cent water. It is a rich source of milk fat, proteins and minerals. The residue obtained from the traditional *makkhan-ghee* process is appreciably richer in proteins and minerals. Its is consumed with rice or leavened bread and also in the preparation of ghee-toffees or *burfi* type sweets.

2

Fermented Milk Products

Fermented milk products, also known as cultured dairy foods, cultured dairy products, or cultured milk products, are dairy foods that have been fermented with lactic acid bacteria such as Lactobacillus, Lactococcus, and Leuconostoc. The fermentation process increases the shelf-life of the product, while enhancing the taste and improving the digestibility of milk. There is evidence that fermented milk products have been produced since around 10,000 BC. A range of different Lactobacilli strains has been grown in laboratories allowing for a wide range of cultured milk products with different tastes.

HISTORY

Fermentation of milk dates back thousands of years to locations all over the world. This process was used as a means of preserving a highly perishable product and to produce new flavors for an old food staple. In the early years of milk fermentation, milk was simply allowed to be fermented by its normal microbiota, but the actual process was not completely understood. Cultures could be maintained by inoculating fresh milk with fermented milk. . Today, lactic acid-producing microorganisms are added to milk to decrease the pH of the milk and produce many different fermented milk products.

PATHWAY

The general pathway for fermentation of milk involves the production of lactic acid from lactose in the milk. This lowers the pH and results in a variety of products. The glucose can be replaced by many different sugars, including lactose, to produce the same product. The protein in milk that is responsible for curdling is casein. With the pH drop, casein molecules coagulate in a structure.

PRODUCTS

Yogurt

Yogurt is produced from whole or skim milk. It is inoculated with a starter culture, which usually contains*Lactobacillus bulgaricus* and *Streptococcus* *thermophilus*. These bacteria ferment the lactose in the milk to lactic acid, causing the milk to curdle and form yogurt.

Sour Cream

Sour cream is produced in much the same way as yogurt. However, light cream is used instead of whole or skim milk. *Streptococcus spp.* and *Lactobacillus spp.* are used as starter cultures to form lactic acid from lactose.

GRAM STAIN OF Streptococcus lactus IN BUTTERMILK, SHOWING CASEIN

Kefir

Kefir is a fermented milk product that utilises a wide variety of microorganisms to produce a wide variety of products in addition to lactic acid including ethanol, free fatty acids, and acetaldehyde . At the base of kefir production are kefir grains. These grains, which are not actually what are traditionally known as grains, contain many microorganisms including bacteria such as *Lactobacillus spp.* and non- pathogenic *Streptococcus spp.*, yeasts such as *Saccharomyces delbrucki* and *S. cerevisiae*, as well as many other organisms that have yet to be classified. When milk is inoculated with these grains, it is allowed to ferment for 18 to 24 hours, during which the variety of products.

Cheese

Species	Major known function	Product
Propionibacterium shermanii	Flavour & eye formation	Swiss cheese family
Lactobacillus bugaricus *Lactobacillus lactis* *Lactobacillus helveticus*	Acid and flavour	Bulgarian buttermilk, yogurt, kefir, koumiss, Swiss, Emmental, and Italian cheeses
Lactobacillus acidophilus	Acid	Acidophilus, buttermilk
Streptococcus thermophilus	Acid	Emmental, Cheddar, and Italian cheeses, and yogurt
Streptococcus diacetilactis	Acid	Sour cream, ripe cream, butter, cheese, buttermilk and starter cultures.
Streptococcus lactis *Streptococcus cremoris*	Acid	Cultured buttermilk, sour cream, cottage cheese, all types of foreign and domestic cheeses, and starter cultures.
Streptococcus durans *Streptococcus faecalis*	Acid and flavour	Soft Italian, cheddar, and some Swiss cheeses.
Leuconostoc citrovorum *Leuconostoc dextranicum*	Flavour	Cultured buttermilk, sour cream, cottage cheese, ripened cream butter, and starter cultures.

Cheese is yet another product of milk fermentation that requires factors in addition to the traditional production of lactic acid. The coagulation of casein due to lactic acid production and the subsequent drop in pH and the addition of rennet, an enzyme derived from the lining of the stomachs of calves, form the curd of cheese Without rennet, a soft cheese such as cottage cheese or cream cheese would result. Yeasts, molds, and bacteria are all involved in the processes that produce different cheeses, and their locations on or in the curd are what results in many cheese types. The bacteria involved in different cheeses:

FUNCTION OF FERMENTED MILK PRODUCTS

The primary function of fermenting milk was, originally, to extend its shelf life. With this came numerous advantages, such as an improved taste and enhanced digestibility of the milk, as well as the manufacture of a wide variety of products. Historically the fermentation of milk can be traced back to around 10 000 B.C. It is likely that fermentation initially arose spontaneously from indigenous microflora found in milk. Fortunately, the bacteria were lactococci and lactobacilli which typically suppress spoilage and pathogenic organisms effectively. The evolution of these products likely came as a result of the climate of the region in which they were produced: thermophilic lactic acid fermentation favours the heat of the sub-tropics; mesophilic lactic acid fermentation occurs at cooler temperatures. Today the fermentations are controlled with specific starter cultures and conditions. Some of the many fermented milk products are: acidophilus milk, crème fraîche, cultured buttermilk, kefir, koumiss, filmjölk, sour cream, and viili. Yogurt and cheese are also fermented milk products. More detail on yogurt and cheese can be found under their specific ingredient profiles.

Fermented milk products can be classified into 3 categories:

- Viscous products
- Beverage products
- Carbonated products.

Within these categories, the fermented milk products may be fresh, or have an extended shelf life. The fresh products contain live starter culture bacteria, including probiotics, while the extended shelf life products contain no live microorganisms.

Product	Typical Shelf Life (4°C)*
Acidophilus Milk	2 wks
Cultured Buttermilk	10 d
Sour Cream	4 wks
Kefir	10-14 d
Koumiss	10-14 d
Filmjölk	10-14 d
Viili	14 d
Crème Fraîche	10 d

**The shelf life may vary with manufacturers.*

There are numerous factors which affect the outcome of the product including the chemical composition of the milk, additives and starter cultures used, as well as the processing of the product. They affect the ultimate flavour, texture, and consistency of the final product. It is not uncommon for the manufacturer to add stabilizers such as pectins and gums, in order to avoid the sedimentation of milks solids and the separation of whey in the package, while improving the mouthfeel of the product.

The general process by which fermented milk products are made begins with a preliminary treatment of milk which may include clarification, fat separation and standardisation, and evaporation. Processing follows next, with de-aeration, homogenisation, and pasteurization. The milk is then cooled to the appropriate fermentation temperature and starter cultures are added.

Starter cultures differ for each product. They consist of microorganisms added to the milk to provide specific characteristics in the finished fermented milk product in a controlled and predictable manner. The primary function of lactic acid starters is to ferment lactose into lactic acid, but they may also contribute to flavour, aroma and alcohol production, while inhibiting spoilage microorganisms. A single strain of bacteria may be added, or a mixture of several microorganisms may be introduced. The bacteria, yeasts and moulds work at different temperatures as well. Thermophilic lactic acid fermentation favour hot temperatures (40-45°C) while mesophilic lactic acid fermentation occurs at cooler temperatures (25 and 40°C).

As the starter cultures grow within the milk, fermentation takes place. Fermentation is the chemical conversion of carbohydrates into alcohols or acids. In fermented milk products both alcohol and lactic acid may be produced, like in kefir and koumiss, or just lactic acid, like in sour cream. The bacteria ingest the lactose (milk sugar), and release lactic acid as waste causing the acidity to increase. This rise in acidity causes the milk proteins to denature (unfold) and tangle themselves into masses (curds) while also inhibiting the growth of other organisms that are not acid tolerant. Following the completion of fermentation, flavourings can be added and the products are packaged, labeled and put into cold storage before being sent to stores.

Characteristics

- Kefir is a milk product traditionally fermented by "kefir grains". The grains are curds which act as a starter culture in each batch of kefir. These grains contain active microorganisms and when added to fresh milk, they produce kefir. Kefir grains have a complex microbial composition consisting of 83-90% lactic acid bacteria and 10-17% yeast, as well as acetic acid bacteria and possibly mould. Commercial starter cultures have been developed that allow production to be made more efficient and may also provide a longer shelf life for the product. The fermentation of the lactose by the microorganisms yields a sour, carbonated, slightly alcoholic beverage with the consistency of thin yogurt. It is white or yellowish in colour with a yeasty aroma. The taste is acidic, but refreshing with compounds including lactic acid, diacetyl, carbon dioxide and ethanol influencing its sensory properties. Kefir is sometimes commercially available without carbonation and alcohol (when yeast is not added to the starter culture), resulting in a product that is very similar to yogurt. Its composition and flavour is dependent on milk type and lactic acid content in the final product. Typical milks used for kefir include cow, goat, and sheep, with each eliciting varying nutritional and sensory qualities.

- Koumiss (koumiss, kumiss, kumis, kymis, kymmyz) is a fermented drink traditionally made from the milk of horses by people in Central Asia and from camel's milk inMongolia . The word *koumiss* is thought to derive from the name of the Turkic Kumyks people. The capital of Kyrgyzstan, Bishkek, is named after the paddle used to churn the fermenting milk, showing the importance of the drink in the national culture. It would have been originally fermented in a horse hide bag which would have contained the microflora from the previous batch. Koumiss is similar to kefir, but is not produced using "grains", but using a liquid starter culture composed of lactobacilli and non-lactose-fermenting yeasts instead. As mare's milk has a higher sugar content than cow's and goat's milk, the resulting koumiss has a slightly higher alcohol content than kefir. Today, cow's milk is generally used for koumiss, with the addition of sugar to better approximate the composition of mare's milk. Koumiss is a milky white liquid with a grayish cast and is very light in body compared to most dairy beverages. It has a slightly sour flavour from lactic acid, and ethyl alcohol, and a fizziness from carbon dioxide.
- Filmjölk is Swedish fermented milk. Fermented milk products have a long history in the Scandinavian countries, dating back to the Vikings. Traditional fermented milks of the Nordic countries are fermented at lower temperatures by mesophilic bacteria (specifically). Filmjölk is the modern variant of the traditional surmjölk. It is the most common fermented milk in Sweden and is frequently consumed for breakfast or lunch. Filmjölk is similar to yogurt or kefir, but it is fermented using different strains of bacteria, giving it a unique flavour. Filmjölk is a spoonable, semi-solid product made with standardized fat contents. It has a mild, slightly acidic flavour, with aromas from diacetyl and carbon dioxide.
- Viili is traditional Finnish fermented milk made from unhomogenized milk. As the milk begins to ferment a layer of cream rises to the surface and a surface growing mould, *Geotrichum candidum,* forms a thin velvety surface. Viili is inoculated with a starter culture containing *Lacobacillus lactis* subsp. *cremoris* that creates its ropy character. It is advised that viili be eaten with a tablespoon so that

it can be cut into portions. If it is mixed or eaten with a teaspoon the texture becomes ropy, making it difficult to consume. Viili has a mild acidic flavour and aroma with a thick consistency that maintains its shape without collapsing when placed on a plate.

- Acidophilus milk is typically a lowfat or nonfat milk to which active cultures of*Lactobacillus acidophilus* have been added. The milk can be refrigerated to prevent further growth of the harmless bacteria producing sweet acidophilus milk. It can also be incubated at 38°C until a curd forms. *Bifidobacterium bifidum* may also be included.
- Cultured Buttermilk has been produced as long as butter has been made. Traditionally butter was made by churning milk or cream, but an improved method for fermenting milk became the preferred method for cultured buttermilk production. Cultured buttermilk is pasteurized skim milk fermented by a lactic acid bacteria culture (*Lactobacillus lactis* subsp. *lactis*, *Lactobacillus lactis* subsp.*cremoris*, and *Lactobacillus lactis* subsp *lactis* biovar. *diacetylactis*, and*Leuconostoc mesenteroides* subsp. *cremoris*) and by aroma bacteria. It possesses a mild acid flavour with a diacetyl overtone and a smooth texture. Cultured buttermilk has a soft white colour and may contain added butterflakes, fruit condiments, or flavourings.
- Sour Cream is an extremely viscous product that has been used for many years in a great number of countries. Traditionally, cream was left to sour, but today sour cream is made by lactic acid fermentation of cream using *Streptococcus lactis*, with or without the addition of rennet to create a thicker product. Stabilizers may be added to improve and maintain the consistency. Sour cream has a mild, subtle, tangy flavour and aroma which is similar to cultured buttermilk. It has a smooth, thick body and typically has a fat content of 10-14%. Lower fat varieties are also produced. Sour cream has a limited shelf life due to yeast and mould growth. The shelf life can be extended by a heat treatment after the fermentation has taken place.
- Crème fraîche is French for fresh cream. It is mild in taste and slightly acidic, with a smooth, rich, thick texture. It is made in the same manner as sour cream, and used for many of the same applications. It is higher in fat content (usually 30-40% fat) and as a result crème fraîche can be whisked into whipped cream. It also has a high enough fat content and low enough protein content that it can be cooked directly without curdling.

Varieties

- Kefir is made most often from partially skimmed cow's milk. It can be packaged either as natural or plain kefir with no added fruit or flavours or as flavoured kefir. The final product contains live bacteria and yeasts that produce carbon dioxide gas. This gas production gives kefir a "sparkling" sensation on the tongue when eaten. Kefir has been referred to as the champagne of fermented dairy products.

- Koumiss: Mare's milk has higher sugar content than cow's and goat's milk, and as a result koumiss has a slightly higher alcohol content than kefir. Today, cow's milk is generally used for koumiss, with the addition of sugar to better approximate the composition of mare's milk.
- Cultured Buttermilk may contain added butterflakes, fruit condiments, or flavourings. It is also available with different fat contents.
- Viili comes in a wide range of varieties, including products of different fat content,lactose-reduced varieties and flavoured versions. Viili can be made from homogenizedmilk and without mould growing on the surface.
- Sour cream comes in full fat (minimum 14% fat), low fat and fat free varieties.
- Filmjölk has fruit flavoured variants and can have the addition of beneficial probiotic bacteria such as *Bifidobacterium lactis* and many species of lactobacilli.

COMPOSITION

Based on	Kefir (100 g)	Sour Cream	
		Low Fat	Fat-Free
		10.6 g	0 g
		3.5 g	3.1 g
		7.1 g	15.6 g
		0.22 g	0.39 g
		141 mg	125 mg

ased on 00 g	Koumiss (mare's milk)	Koumiss (cow's milk)
Fat	1.1 g	3.9 g
otein	3.5 g	3.3 g
hydrates	6.1 g	4.7 g
ugar	6.1 g	4.7 g
lcium	90 mg	~120 mg

Based on 100 g	Cultured Buttermilk		Acidophilus milk (2%)	Crème Fraîche
	Low Fat	Fat Free		
Fat	2.0 g	0.88 g	3.8 g	38.7 g
Protein	4.1 g	3.3 g	1 g	1.8 g
Carbohydrates	5.3 g	4.8 g	5.4 g	2.5 g
Sugar	5.3 g	4.8 g	5 g	2.5 g
Calcium	143 mg	116 mg	125 mg	60 mg

Viili and filmjölk likely have compositions similar to kefir depending on the level of milk fat used to prepare them.

VARIOUS USES

Kefir and Koumiss can be used in smoothies, salad dressings, and sauces. They can be added to baked goods such as pancakes, waffles, and breads, or in soups and desserts as a replacement for other milk products such as yogurt or buttermilk. They are also delicious mixed with fresh fruit or cereal as a breakfast, lunch or snack. They all make refreshing beverages on their own or mixed with fruits, honey, maple syrup, iced coffees and teas as well as other sweeteners and flavours

Filmjölk is eaten in the same way as yogurt, usually from a bowl using a spoon. It is sometimes drunk as a thick beverage. Many people add sugar, jam, applesauce, cinnamon or berries. Cereals, corn flakes or muesli are often added to filmjölk. In northern regions of Sweden, crushed crisp bread is sometimes put into it. It could be used in smoothies, salad dressings and sauces, as well as in baked goods.

Crème Fraîche has similar uses to sour cream, however it's sweeter flavour makes it particularly well suited to desserts, as a topping, or as a base for other flavours. Crème fraîche works well in dips, dressings and sauces or as an addition to soups.

Sour Cream has numerous applications. It is commonly used as a base for dips, salad dressings and sauces. It is eaten as a condiment on potatoes, chili, or with smoked salmon, as well as many other foods. Sour cream can be used in soups and works well in baked products like breads, cakes, pies and cookies. Sour cream has significantly less calories than mayonnaise and performs many similar functions. In Russian cuisine, sour cream is often added to borscht and other soups. In Tex-Mex cuisine, it is often added to tacos, nachos, burritos, taquitos or guacamole. Hungarian cooks use it as an ingredient in sauces and in recipes such as ham-filled crepes.

Cultured Buttermilk is a versatile ingredient in baking. It works very well in biscuits, breads, and desserts. Cultured buttermilk is often used in salad dressings and sauces, stirred into mashed potatoes and soups and it has even been used to make tangy buttermilk ice cream. It is also considered a refreshing beverage. Viili is consumed fresh and chilled, in the same way as yogurt, and can be topped with fruits, nuts, or cereal, as well as other flavourings like spices, sugar and honey. Viili may be added to smoothies, or used in baked goods. It may also be flavoured like yogurt. Acidophilus milk is consumed as a beverage by the glass, or added to cereal. It has been used to make egg nog. Acidophilus milk also works well in sauces and desserts.

FUNCTIONAL PROPERTIES

Fermented milk products have numerous functional properties:

- Preservation: bacteria are inhibited from growing through pH reduction when lactic acid is formed, and shelf life is increased

- Flavour Enhancement: the sour characteristic of fermented milk products comes from fermentation products (lactic acid, diactyl, carbon dioxide, ethanol); these products act as excellent flavour carriers for herbs, spices and other flavourings
- Texture Enhancement: some fermented milk products (sour cream or crème fraîche) can add body and thickness to sauces, dips or vinagrettes
- Reducing Caloric Content: many fermented milk products come in low fat or fat free varieties and can be used to substitute for higher fat ingredients
- Emulsification: milk proteins help stabilize fat emulsions in salad dressings, soups and cakes
- Foaming and Whipping: crème fraîche is capable of being whipped like whip cream
- Nutritional benefits: fermented milk products may contain probiotics (bacteria that are beneficial to health) as well as many vitamins and minerals.

PRODUCTION OF CULTURED MILK

Under unrefrigerated storage, milk becomes sour in a matter of hours after milking. This is due to the action of natural lactic acid bacteria on the milk sugar, turning it to lactic acid which causes the milk protein to coagulate. Traditionally, most communities in Africa and Kenya in particular have used this natural fermentation to produce a variety of fermented milks for home consumption.

Under modern dairy processing industry, selected lactic starter cultures are used to ferment milk during preparation of a variety of cultured dairy products. Since, the liberalisation of the dairy industry in Kenya in 1992, a number of small to medium scale dairy processing plants have emerged. Apart from processing liquid milk, most of these plants are also producing a variety of cultured milk products. The quality of these products vary from being of average or below average quality to high quality products. In order to assist the small to medium scale processors improve the quality of their products, the Training Programme for the Small Scale Dairy Sector under project GOK/FAO/TCP/KEN/6611, has prepared this guide on Production of Cultured Milk to he used for training and by the private small scale dairy processors. The emphasis is on hygienic milk production methods and a good understanding of starter culture technology and handling method which is essential for successful production of good quality cultured dairy product.

PRODUCTION OF CULTURED MILK

Cultured milks are products made by use of special lactic acid bacteria cultures. They fall into two broad categories. Those made by use of lactic acid bacteria which grow well at ambient temperature (25-30°C). Such lactic acid bacteria are known as Mesophilic starter cultures. Maziwa lala or Mala as it is popularly known in Kenya, is made by use of such cultures. The other type of cultured dairy product is the one made by use of lactic acid bacteria which grow well under warm conditions (38 - 45(C) The lactic acid bacteria used are technically known as Thermophilic starter cultures. Yoghurt or yoghurt like products belong to this group.

TYPES OF CULTURED MILK AND STARTER CULTURES

The cultured dairy products fall in two broad categories; those made by use of mesophilic starter cultures and those made by use of thermophilic starter cultures. Depending on the type of culture used, the flavour, texture and consistency can vary widely and may be grouped as follows:

Products made by use of mesophilic lactic starter cultures may use one of the following starter culture types:

O-type: These are starter cultures in which the main lactic acid bacteria are Lactococcus lactis subsp.lactis and Lactococcus lactis subsp. cremoris. These produce mainly lactic acid. They are thus, homofermentative.

D-type: These are starter cultures containing, in addition to the O-type bacteria also a flavour producing lactic bacteria known as Streptococcus lactis subsp. lactis var. diacetylactis. This, as the name implies, produces a flavour compound known as diacetyl which gives a flavour characteristic of cultured cream butter or its buttermilk by-product. In addition to production of diacetyl, it also produces carbon dioxide which contributes to the blend of a delicate flavour. L-type: This type of starter culture contain, in addition to the O-type bacteria, also Leuconostoc mesenteroides subsp. mesenteroides as the main flavour compound producing bacteria. It produces diacetyl, acetic acid, acetaldehyde and other flavour compounds but less carbon dioxide than the D-type. LD-type: These contain a combination of Str. lactis subsp. lactis var diacetylactis and Leuconostoc mesenteroides subsp. mesenteroides to give a fine blend of dedicated flavour and aroma.

Most "Maziwa lala" products can be made by use of any of these types of lactic starter cultures. They will all give products differing in flavour and texture. Which ever one chooses to use, care should be taken not to use those that lack the desired flavour or those which halve excessive gas production (good for some cheese, but not entirely welcome in good cultured dairy products!). Excessive production of acetaldehyde can also lead to a defect known as "green flavour" more akin to yoghurt than to "Maziwa lala" group of cultured products.

Products made by use of thermophilic starter cultures.

The main product in this group is yoghurt. The typical lactic acid bacteria in yoghurt starter culture are Streptococcus salivaricus subsp. thermophilus and Lactobacillus delbrueckii subsp. bulgaricus. The former is responsible for fermenting lactose to lactic acid whereas Lb. delbrueckii subsp. bulgaricus is responsible for flavour production, mainly in the form of acetaldehyde. Due to its proven health food qualities, other thermophilic lactic acid bacteria may be included in yoghurt starter cultures to enhance its dietetic and health food status. Lactobacillus acidophilus, which produces a natural antibiotic, acidophilin and Bifidobacterium spp., which are part of the natural bacterial flora of the human gut, are often included in special yoghurt starter cultures. Products such as Bioghurt are made by use of such specially blended cultures and are marketed under the banner of "Health foods" useful for the restoration of normal bacterial flora hollowing a course of oral antibiotic therapy in patients.

WHY MAKE CULTURED MILK?

Apart from fresh milk, fermented milk is the most common milk food in Africa and Kenya in particular: There is thus, a big demand for cultured milk which is perhaps not yet well exploited. In general, the production of cultured milk by dairy processors in Kenya should be advocated further on the following grounds:

- Cultured milk is a milk product with a shelf life (at ambient temperature) of about 4 days compared to 24 hours for flesh pasteurised milk.
- Cultured milk is highly digestible making it to have, on a volume to volume basis, better nutritional quality than liquid milk.
- Certain individuals who do not consume milk regularly experience stomach upsets if they try to consume a glass or two of fresh milk. This is because they are unable to digest the milk sugar (lactose) which is then fermented in the lower gut thereby causing gas production and flatulence. Experience has shown that such, so called "lactose intolerant" people can consume cultured milk without problems because a large part of the milk sugar has already been digested for them by the lactic acid bacteria!
- The regular consumption of cultured milk has been shown to have therapeutic properties due partly to the presence of natural antibiotics such as nisin and acidophilin which is produced by specific lactic acid bacteria. This has gives certain cultured products such as Acidophilus milk, yoghurt (Bioghurts) special "Health food) status and may be recommended by medics for restoration of normal gut flora following oral therapy.

CHOOSING TYPE OF CULTURED PRODUCT

Most dairy processors in Kenya are making either "Maziwa lala" type of yoghurt type of cultured milk. Whichever product one chooses to manufacture, it is important to consider appropriateness and desirability.

"Maziwa lala" type products. These have the advantage in that their flavour is close to that of naturally fermented milk to which most of the Kenyan cultured milk consuming public is familiar with. Furthermore fermentation takes place at ambient temperatures obviating the need for incubators. Even without cooling of the coagulum, the fermentation process stops at 0.85-1.0% Lactic acid (pH 4.3 - 4.2) resulting in mild sour products with a shelf life of up to 4 days without refrigeration.

Yoghurt and yoghurt like products. Yoghurt is a well liked dairy product in Europe and North America, where sales have increased substantially during the last two decades partly because of its health food aspect and its ability to blend so well with a variety of exotic tropical fruits/flavours. It requires incubator at 42°C during manufacture and has to be cooled to stop further acidification to less than pH 4.0 - 3.8. It thus costs more to process than its "Maziwa lala" counterpart.

In view of this, the production and marketing of yoghurt in Kenya is normally targeted at the upper niche of the market segment in metropolitan cities of Nairobi and Mombasa. Demand is likely to increase with increase in peoples' incomes. With so much bananas. pineapples, papayas, mangoes, oranges, passion fruits, straw berries around us in Kenya, the opportunity to produce yoghurt fruit blends is not yet fully exploited!

Table. Summary of Starter Cultures and Cultured Milk

SOURCING FOR STARTER CULTURES

There are a few specialised starter culture suppliers. Commercial starter culture suppliers such as Chr. Hansen1 supply three types of starter cultures:

Dri-vac: This type contains inactive microorganisms in freeze dried term. They must be activated by propagation at least through 3 passages before they can be used in the production of cultured milk product.

Redi-set: This type has active lactic acid bacteria and is used directly in the production of cultured milk products. It has the advantage of avoiding possible contamination associated with serial transfers of starter cultures during propagation. It also enables the production of products of uniform quality throughout the production year. Its main disadvantage is that it is considerably more expensive than the Dri-vac type. DVS: This is similar to Redi-set but stronger (more concentrated) and has the same advantages and disadvantages.

STARTER CULTURE PROPAGATION

Preparation of starter culture and maintenance of its quality is an important aspect of production of cultured milk. Once you obtain a new supply of freeze dried starter culture proceed as follows:

1. First Passage
   a. Obtain plastic bottles that hold 880 - 1,200 ml of liquid and have screw type covers.
   b. Select good quality raw milk.
   c. Skim the raw milk using a milk separator (skim milk has best results, however one can use whole milk).
   d. Sterilize skimmed milk by heating to 100 °C for 10 - 20 minutes, or 90 - 95°C for 30 minutes constantly stirring to ensure even heating.
   e. Wash and sterilise the plastic bottles and lids, either by boiling for 30 minutes or by disinfecting in 68 -70% (vol) ethyl alcohol then place them upside down on a sterile towel to dry. Sterilize the towel by moistening it with alcohol.
   f. Cool the milk to room temperature, 20 - 25°C or 30 - 37°C in the case of yoghurt starter culture.
   g. After the milk has cooled, add 1 to 2 grains of freeze - dried culture per litre of sterilized skimmed milk and stir for one minute.
   h. Using only sterile equipment, fill the plastic bottles about two - thirds full and immediately screw on covers.
   i. Allow the inoculated milk to incubate at room temperature for 16-24hours or 42 °C for aft hours in the case of yoghurt starter.

1Technoserve, 1994. Male Milk Manual; A guide for establishing and operating small scale Enterprises for production of Cultured Milk: (Technoserve- P. O. Box 14821 Nairobi: or P.O. Box 2117. Arusha, Tanzania).

2. Second Passage.
   a. Repeat First Passage Steps a) to e) using a litre of milk.
   b. Pour the hot skimmed milk into the bottle to two-thirds its capacity cover tightly.
   c. After the bottled sterilised skim milk has cooled to room temperature, add the First passage inoculum into each bottle in a 2 - 3% mixture by volume and stir.

Close each bottle tightly and leave it to incubate fur 16 - 24 hours (or 3 - 4 hours at 42 °C in the case of yoghurt) to produce inoculum for the next passage.

d. Discard any remaining First Passage inoculum.

3. Third passage.
   a. Follow second passage procedures to produce the third passage culture inoculum. This active mother culture may be used for the production of bulk starter to be used in the production of cultured milk ("Maziwa lala" or Yoghurt). Keep mother culture in a refrigerator for not more than 5 days, then re-activate by repeating 2nd passage procedure.

PRESERVATION OF MOTHER CULTURE BY DEEP-FREEZING

In order to cut down the cost of obtaining fresh freeze dried starter cultures each time one wants to prepare a fresh stock of mother culture, it is possible to preserve some of the mother culture obtained after the 3rd Passage above.

Do as follows:

a. Prepare 20 - 30 plastic bottles of 100 - 250 ml capacity.
b. Prepare good quality skim milk and fill the plastic bottles to 2/3 of their capacity.
c. Sterilise as described.
d. Cool to ambient temperature.
e. Inoculate with mother culture at the rate of 2 - 3%, close the bottles and immediately place in a deep freezer.(After inoculation the acidity of the sterilised milk should not be less than pH 6.4). Such culture will remain viable for up to 1- 2 year if not allowed to thaw (defrost).

PREPARATION OF STARTER CULTURE FRONT FROZEN FLUID CULTURE

i. Remove from the deep-freezer one of the frozen liquid starter cultures, when required.
ii. Thaw at room temperature and leave to coagulate in 16 - 24 hour) (for yoghurt incubate at 30 °C for 16 - 20 hours).
iii. Select good quality raw whole milk.
iv. Heat the milk to 90 - 92 °C for 10 - 20 minutes, stirring constantly to avoid cream separation and ensure even heating.
v. Sterilize plastic bottles (100 - 1,500 capacity), with screw type lids and place them upside down on a sterile towel to dry.
vi. Fill the bottles two-thirds full with the hot milk and cover them tightly. Let them cool at room temperature for 20 - 30 minute.
vii. Place the bottles in water bath to cool to 20 - 25 °C (or 42 °C in the case of yoghurt starter).
viii. Inoculate die sterilised whole milk in bottles with 2 - 3 % of the coagulum produced.
ix. Make two more transfers to activate the starter culture in the same.

x. Use coagulated inoculum as mother culture for the production of bulk starter to be used in the production of your cultured milk. Each time your bulk mother culture becomes exhausted or weak, go for another deep-frozen liquid culture and repeat the above procedure. This procedure wilt enable you to extend the use of one original pouch of freeze dried culture for up to 2 - 3 years!

BUTTERMILK

Buttermilk refers to a number of dairy drinks. Originally, buttermilk was the liquid left behind after churning butter out of cream. This type of buttermilk is known as *traditional buttermilk.*

The term *buttermilk* also refers to a range of fermented milk drinks, common in warm climates (e.g., the Balkans, the Middle East, Turkey, Afghanistan, Pakistan, Nepal, India, Sri Lanka, Nicaragua and the Southern United States) where unrefrigerated fresh milk sours quickly, as well as in colder climates, such as Scandinavia, Ireland, Netherlands, Germany, Poland, Slovakia, Slovenia, Croatia and the Czech Republic. This fermented dairy product known as *cultured buttermilk* is produced from cow's milk and has a characteristically sour taste caused by lactic acid bacteria. This variant is made using one of two species of bacteria—either*Lactococcus lactis* or *Lactobacillus bulgaricus*, which creates more tartness.

The tartness of buttermilk is due to acid in the milk. The increased acidity is primarily due to lactic acid produced by lactic acid bacteria while fermenting lactose, the primary sugar in milk. As the bacteria produce lactic acid, the pH of the milk decreases and casein, the primary milk protein, precipitates, causing the curdling or clabbering of milk. This process makes buttermilk thicker than plain milk. While both traditional and cultured buttermilk contain lactic acid, traditional buttermilk tends to be less viscous, whereas cultured buttermilk is more viscous.

Buttermilk can be drunk straight, and it can also be used in cooking. Soda bread is a bread in which the acid in buttermilk reacts with the rising agent, sodium bicarbonate, to produce carbon dioxide which acts as the leavening agent.

Traditional buttermilk

Originally, buttermilk referred to the liquid left over from churning butter from cultured or fermented cream. Traditionally, before cream could be skimmed from whole milk, the milk was left to sit for a period of time to allow the cream and milk to separate. During this time, naturally occurring lactic acid-producing bacteria in the milk fermented it. This facilitates the butter churning process, since fat from cream with a lower pH coalesces more readily than that of fresh cream. The acidic environment also helps prevent potentially harmful microorganisms from growing, increasing shelf-life. However, in establishments that used cream separators, the cream was hardly acidic at all. On the Indian subcontinent, the term "buttermilk" refers to the liquid left over after extracting butter from churned yogurt. Today, this is called *traditional buttermilk*. Traditional buttermilk is still common in many Nepalese, Indian and Pakistani households but rarely found in western countries. In Southern India and most areas of the Punjab,Saurashtra (Gujarat), buttermilk with added water, sugar or salt, asafoetida, and curry leaves is a must-have in daily food while also given at stalls in festival times.

Health benefits

Buttermilk prepared in the traditional way is considered beneficial to health as it contains probiotic microbes and is sometimes referred to as "Grandma's probiotic". It is also soothing to stomach and skin. The fat content of buttermilk is far lower than that of milk or curd as fat is removed during churning. The probiotic nature of buttermilk is purported to be beneficial to the gut and improve immunity when taken regularly. One cup of whole milk contains 157 calories and 8.9 grams of fat whereas one cup of buttermilk contains 99 calories and 2.2 grams of fat. Buttermilk contains vitamins, potassium, calcium, and traces of phosphorus. In some countries, such as India, it is a favourite traditional drink during summer as it is soothing to the stomach and alleviates minor stomach upsets. In India, flavoring ingredients such as asafoetida, coriander leaves, ginger, curry leaves and sea salt are mixed with buttermilk to enhance its digestion-aiding properties.

Cultured buttermilk

Commercially available cultured buttermilk is milk that has been pasteurized and homogenized (with 1% or 2% fat), and then inoculated with a culture of Lactococcus lactis(formerly known as Streptococcus lactis) plus Leuconostoc citrovorum to simulate the naturally occurring bacteria in the old-fashioned product. Some dairies add colored flecks of butter to cultured buttermilk to simulate residual flecks of butter that can be left over from the churning process of traditional buttermilk. Condensed buttermilk and dried buttermilk have increased in importance in the food industry. Buttermilk solids are used in ice cream manufacture, as well as being added topancake mixes. Adding specific strains of bacteria to pasteurized milk allows more consistent production. In the early 1900s, cultured buttermilk was labeled *artificial buttermilk*, to differentiate it from traditional buttermilk, which was known as *natural* or *ordinary buttermilk*.

Acidified buttermilk

Acidified buttermilk is a related product made by adding a food-grade acid (such as lemon juice) to milk. It can be produced by mixing 1 tablespoon of vinegar or lemon juice with 1 cup of milk and letting it sit until it curdles, about

10 minutes. Any level of fat content for the milk ingredient may be used, but whole milk is usually used for baking. In the process which is used to produce paneer such acidification is done in the presence of heat.

Powdered buttermilk

Like powdered milk, buttermilk is available in a dried powder form. This stores well at room temperature and is usually used in baked goods.

PREPARATION OF BULK STARTER FROM MOTHER CULTURE

a. Follow steps (iii) to (vii) as if preparing mother culture from frozen liquid culture.
b. After cooling to the appropriate incubation temperature, inoculate with 2-3% of fluid mother culture and incubate until coagulation has occurred.

Note: Preparation of bulk starter culture need to be one every day from the same mother culture.

A change of the mother culture from which the bulk starter culture is prepared daily will only become necessary if and when:

- Coagulation of milk takes longer than usual in inhibitor free milk.
- The final product develops off-flavours or defects such as excessive gas yeasty taste and flavour etc.
- A mishandling or contamination of the mother culture has occurred.

Always ensure that enough bulk starter culture is available for the next day's production i.e 10 - 20 litres for every 500 litres of cultured milk.

PRODUCTION OF CULTURED "Maziwa lala"

Cultured milk way be produced in quantities ranging from as little as 50 litre per day to 500 litres per day for small scale dairy plants to several thousand litres per day in medium scale dairy plants. Whatever the size of your operation, the basic steps are the same. The difference will be in the type and size of equipment used. We shall only discuss processes involved. For further information of suitable equipment for the size of operation envisaged contact the Dairy Training Institute at Naivasha or the Department of Food and Dairy Technology at Egerton University and Department of Food Technology and Nutrition University of Nairobi at Kabete

1. For production of good quality "Maziwa lala" you require good quality milk
2. Ensure you have good starter culture.
3. Filter the milk to remove physical dirt.
4. Pasteurise the milk at 80 - 85(C for 30 minutes in a batch pasteuriser which can be a 50 litre can heated in hot water or improved wood fueled jiko.

This is done to:

- Destroy spoilage and pathogenic bacteria.
- Produce a smooth and viscous product.
- Prevent wheying off in the sour curd.
- Enable starter cultures to grow well in the milk.

If you are using continuous flow pasteurisers you may heat the milk to 90 - 95 °C for 2 - 5 minutes.

5. After pasteurisation the milk should be cooled to 20 - 25 °C. Cooling may be done rapidly in a cold water basin or trough.

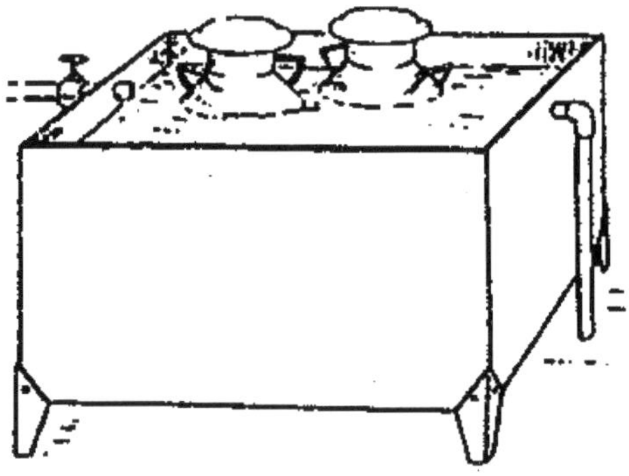

Fig. Cooling in water ensures faster cooling than cooling in air.

6. After inoculate the milk with 2 - 3 % active starter culture. Stir well for 3-5 minutes to ensure uniform distribution of starter culture.
7. Incubate at ambient temperature overnight (for 16 -18 hours).
8. Check the coagulum. It should be compact without cracks and whey on top. You may cool the coagulum before stirring or breaking the curd if you have the facilities. This will improve the viscosity and the consistency of the product but it is not essential.
9. Break the coagulum and stir well to homogenise the curd to a smooth consistency. You may add sugar and flavourings at this stage. To minimise contamination sugar may be added towards the end of the pasteurisation process i.e when the milk is still hot.
10. Pack the finished product in tail containers for marketing. Yon may use plastic sachets (1/2litre to 1 litre or plastic cups 150 ml to 200 ml) Avoid contaminating the product during packaging (swollen packages in the market is often due to contamination at this stage). Ensure the packages are well labelled and attractive.

PRODUCTION OF YOGHURT

The manufacturing process for yoghurt is similar to "Maziwa lala" " with the exception that a yoghurt starter culture is used instead and the milk is incubated under warm conditions. To produce good quality yoghurt proceed as follows:

1. For Production of good quality yoghurt, you require good quality milk.
2. Ensure you have good starter culture.
3. Filter the milk to remove physical dirt.
4. Pasteurise the milk at 80 - 85°C for 30 minutes in a batch pasteuriser which can be a 50 litre can heated in hot water or improved wood fuelled jiko.
5. After pasteurisation the milk should be cooled to 40 - 45°C. Cooling may be done rapidly in a cold water basin or trough.
6. After cooling, inoculate the milk with 2 - 3% active starter culture. Stir well for 3-5 minutes to ensure uniform distribution of starter culture.
7. Incubate at 40-45 °C for a period of 2 - 3 hours.
8. Check the coagulum. It should be compact without cracks and whey on top.
9. Check the acidity. It should be 0.8 - 0.9% lactic acid or pH 4.3 - 4.2.
10. Cool the coagulum by placing cans in chilled water or store in a refrigerator for 12hrs. This is essential in preventing over souring of yoghurt. It also improves consistency and viscosity as well as full yoghurt flavour development.
11. Break the coagulum and stir well to homogenise the curd to a smooth consistency.
12. You may add sugar and flavourings at this stage. To minimise contamination sugar may be added towards the end of the pasteurisation process i.e when the milk is still hot.
13. Pack the finished product in retail containers for marketing. You may use plastic sachets (1/2 litre to 1 litre or plastic cups 150 mls to 200 mls). Avoid contaminating the product during packaging- swollen packages in the market is often due to contamination at this stage. Ensure the packages are well labelled and attractive!

Basic Practical Requirements for Production of Good Quality Cultured Milk:

a. Use high quality milk.
b. Use colostrum and mastitis free milk.
c. Use milk free of starter culture inhibition (chemical and drug residues). Conduct the inhibitor test regularly on milk destined for production of cultured milk.
d. Use clean equipment.
e. Pasteurise the milk to recommended temperatures.
f. Use flesh starter cultures.
g. Cool yoghurt promptly and maintain cold chain during marketing to avoid over-souring; keep "Maziwa lala" cool although this is not as critical as it is for yoghurt.
h. Package the products in attractive, well-labelled containers.

STARTER CULTURES FOR PRODUCTION OF FERMENTED MILK PRODUCTS

A large collection of natural isolates of lactic acid bacteria (LAB) was made mostly from different homemade cheeses and fermented milk products that are produced in very specific ecologically isolated regions throughout Serbia. These regions such as, for instance, the high mountains, the river valleys, the seacoast, etc. are characterised by extremely clean surrounding. In addition, people inhabiting some of these regions are long living population. All strains have been biochemically and genetically characterised in the Institute for Molecular Genetics and Genetic Engineering, Belgrade, Serbia. On the basis of this research, the various starter cultures containing precisely characterised LAB were constructed for production of cheeses and fermented milk products. These starter cultures were tested on industrial scale in dairy plants in Serbia.

ACIDOVIT

PRODUCTION OF "ACIDOVIT"

"Acidovit" is low fat, dietary fermented milk product made from the skimmed pasteurized milk. Manufacturing of "Acidovit" is possible in the standard yoghurt production line without any substantial changes or significant investments. For production of "Acidovit", the only addition to the yoghurt production line is a vat with cooling system.

The final product is dietary fermented milk beverage with 0.1% milk fat, 0.5% of minerals, 3.9% of protein and 5% of lactose (total dry mater of 9.5%). Energy value of "Acidovit" is 151.58 kJ/100gr or 312.57 kJ per 0.2 litre package. Except the good nutritive characteristics typical for all fermented milk beverages, "Acidovit" also has some additional properties because of the presence of *Lactobacillus helveticus*. It has strong antimicrobial activity and inhibits the growth of pathogenic and food spoilage microorganisms, such as bacteria from genus *Escherichia, Proteus, Pseudomonas, Bacillus* and *Clostridium*. "Acidovit" could be manufactured in two versions, according to the taste of consumers.

(a) "Acidovit" from pasteurized skim milk with the strain *Lactobacillus helveticus* BGRA43. After the preparation of milk and starter culture by the procedure similar to that for yogurt production, the milk is inoculated with 2-3% of the culture, and fermentation is performed at 350C within 6 hours. During the fermentation, after the formation of curd (4-4.5 hours, 34-360SH), the fermentation vat is cooled with ice-cold water. At the end of fermentation, a gentle agitation of the curd (15-30 rpm) is recommended. The final fermented product is characterised by the high viscosity and pleasant thickness having a good flavour, and with shelf life of 5 days at 40C.

(b) "Acidovit" from pasteurized skim milk with mild flavour. In this case, the mixed starter culture consisting of strains*Lactobacillus helveticus* BGRA43 and *Streptococcus thermophilus* BGTMS1 is used under the same manufacturing conditions.

YOGHURT

The starter culture for yogurt production (YOGO) is recently constructed from strains *Lactobacillus bulgaricus* BGPKM1 and *Streptococcus thermophilus* BGTMS2. Using this starter culture in the standard yoghurt production line, the final product (obtained after 2.5 to 3 hours of fermentation) had a generally good quality with appropriate viscosity and flavour.

WHITE SOFT CHEESE

Starter culture for production of white soft cheese (*Lactococcus lactis, Lactococcus lactis* subsp. *diacetylactis*) could be added as 3-5% inoculum during the initial steps of cheese making. This starter culture has the aromogenic properties and the very high level of salt tolerance, which enables its activity in the presence of higher concentrations of salt. The use of this culture in cheese making process resulted in much better final product with good flavour and less variations in overall quality.

SEMIHARD TRAPIST-TYPE CHEESE

Starter culture for production of semi hard cheeses using Trapist technology is mesophilic starter culture consisting of two species from genus *Lactococcus* and one *Lastobacillus casei* strain. Preparation of bulk starter was completed within 8 hours using 3-5% inoculum. The ripening period of the cheese was from 2 to 3 months. This starter culture was used on industrial scale in the processing of 3000 L of milk and compared with the activity of imported starter cultures. The cheese manufactured with our starter culture has very good flavour and the elasticity. In addition, during repeated cheese making with our starter culture, obtained cheese always had standard characteristics with minor variations.

TRADITIONAL FERMENTED MILK AND THE DAIRY INDUSTRY

In the traditional method of producing *kivuguto,* once the cow has been milked, the milk is placed in a jar called an *inkongoro*. After this, the milk is poured into a big, clean, wooden jar called an *icyansi* and is left at room temperature in a warm and clean place called an *uruhimbi*. In order to keep the product safe, the jar is usually covered either with a straw-woven lid known as an *umutemeri* or with a lid made from a calabash. A fermentation period of at least 2 to 3 days is required in order to obtain good *kivuguto*. The flow chart for the traditional processing of *kivuguto* and its by-products. The typical flora that creates the traditional sour milk, *kivuguto*, was not, to date, specifically selected and applied in producing the drink in dairy. This flora is very complex and varies from one location to another, as it is the case for all traditional curds. In addition, the lactic flora lives alongside a pathogenic flora, which also varies depending on the level of the personal hygiene of those preparing the *kivuguto*. Nevertheless, despite the disadvantages of the process, traditional fermentation remains the most convenient method in Rwanda. The microorganisms used are indigenous strains, which are left to develop naturally during the production process. However, in order to standardize this process, the need has arisen for the selection and application of specific strains.

These microorganisms are responsible for the acidification, texture, flavour and other organoleptic characteristics, which give the *kivuguto* curd its highly appreciated taste. This microflora is found in containers commonly used for storing milk. Its habitat is also easily located in the near environment of cattle (pasture, paddock, cattle dander, air, etc.).

The preparation of the *kivuguto* curd varies between regions. In most cases, raw milk is stored directly in a vessel until it ferments spontaneously *via* microorganisms contained in the milk. These microorganisms come either from inside container or from the air in the surrounding environment. Note that in all methods of *kivuguto* production, the vessel is not hermetically sealed. Also, from time to time, the *kivuguto* of a previous batch is used in small quantity as a starter added into new fresh milk. In other cases, raw milk is first heated to boiling point and then cooled to room temperature. It is then stored for 24-36 h before coagulation.

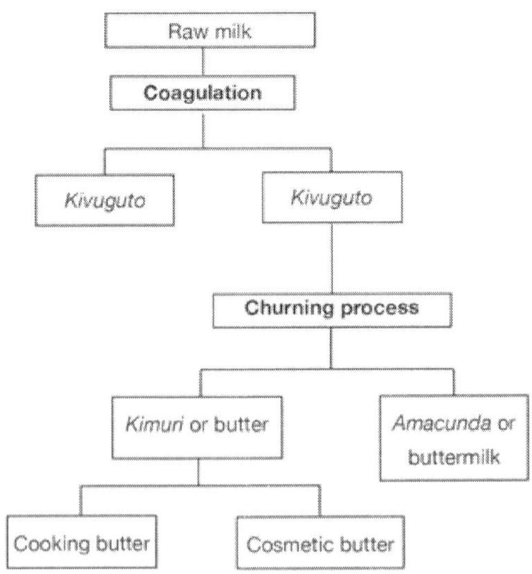

Figure 1. Diagram showing the processing of *kivuguto* and its by-products — *Schéma de préparation du* kivuguto *et de ses dérivés*.

In areas of eastern Rwanda, the containers in which the milk is stored undergo special preparation: washing with the hot smoke derived from burning two species of plant, one of which is an herb of the Poaceae (grass) family, and the other a shrub called *Combretum molle*. The grass used is first stripped of its roots and all its leaves, leaving a single short stem, which is then dried in the sun. These dried stems (rods) are then collected into small bundles and are stored for more than six months in the house. During this time, smoke from the kitchen covers the bundles of rods with a reddish layer. At the end of this period, these rods are burned in a small pot fitted with one large and one small opening.

First, the rods are inserted into the large opening of this small pot and are burnt, second the mouth of a vessel – a small wooden vessel (*inkongoro*), or a small gourd – is placed over the smaller opening of the pot in order to capture all the smoke that emerges. After a while, the vessel is removed and the inside is wiped with a kind of cloth made out of tree bark.

The vessel is then placed back over the emerging smoke, then removed and wiped inside again, and so on. The exercise is repeated several times in order to make the inside of the vessel very smooth. At the end of the procedure, the vessel is kept closed. When milk is placed in a container prepared in this way, the fermentation process takes about 36 h, and a very good curd is produced, with a solid texture. For larger containers, the washing process is carried out using smoke obtained by burning a single large stem of the shrub, which has not been completely dried. The stem is burned and the vessel is held upside down directly over the smoking stem. The vessel is then wiped several times as before. In both procedures, the fermented milk stored in these containers has a smell and taste of smoke, which is highly appreciated both by Rwandese people and by others in the region. What is clear is that this smoking process ensures that these containers are clean, so that there are usually very few pathogens affecting the texture of the curd.

THE TRADITIONAL CHURNING PROCESS

Milk stored in a small gourd will ferment to give *kivuguto* milk, which is then used in this liquid form. On the other hand, when milk is left to ferment in a large gourd (*igisabo*), or in a calabash, additional processing is applied to the resulting *kivuguto* milk. Here the *kivuguto* is churned to produce *kimuri* (butter) or*amacunda* (buttermilk). In order to produce these products, the *kivuguto* is first of all stirred while it bubbles in a butter churn usually in the form of the calabash in which the milk was fermented. This churning process takes at least 2 h. During this time, the fat from the *kivuguto* coalesces and the separation between liquid and solids becomes more marked.

The fat is removed and the liquid left behind in the churn is buttermilk, a product drunk as such. Buttermilk produced in this way has a characteristically sour taste caused by lactic acid bacteria involved in milk lactose hydrolysis during fermentation. The extracted fat is then used in one of two ways. It may be heated with some natural perfuming additives to be used as a cosmetic body cream. On the other hand, the fat may be well preserved for a half-year or a full-year period, allowing the manufacture of a solid tasting yellow fat called butter or *amavuta akuze*, which is good for food preparation. Sometimes, this butter is seasoned with additional ingredients such as onions, garlic, etc., in order to add flavour to meals.

Characterisation of kivuguto

Microflora

We investigated the characteristics of *kivuguto*. Using two samples of *kivuguto* from northern and southern provinces of Rwanda, we selected the microorganisms responsible for the fermentation of the product and formulated a starter culture composed of three strains: CWBI-B1466*Lactococcus lactis* registered in the Genbank database under accession number JF313446, CWBI-B1465 *Leuconostoc mesenteroides* subsp. *mesenteroides*, accession number JF313445 and CWBI-B1470 *Leuconostoc pseudomesenteroides*, accession number JF313454. *Kivuguto* has a lot of similarities with *kule naoto*, a traditional fermented milk of Kenya.

The dominant lactic acid bacteria of *kule naoto* belong to the genus *Lactobacillus*, followed by *Enterococcus*, *Lactococcus* and *Leuconostoc*. *Kivuguto* also presents some similarities with *urubu*, a traditional fermented milk of Burundi . To date, the microorganisms of *urubu* have not been selected. Isono et al. (1994) found that the major microorganisms of the Masai fermented milk in Northern Tanzania consist of*Lactococcus lactis* ssp. *lactis* and *Lactobacillus confus*. Since, 1992, a similar study was carried out on traditional fermented milk of Zimbabwe, *amasi*. Assays for formulating a starter culture for the production of *amasi* were conducted with isolates of *Lactococcus lactis* subsp. *lactis* biovar *lactis*, *Lactobacillus plantarum* and *Leuconostoc mesenteroides* subsp. *mesenteroides*.

However, a sensory panel was subsequently found to prefer *amasi* produced by a single strain of *Lactococcus* . *Lactococcus lactis* is the strain responsible for the fermentation of an israeli milk called *zivda*. The strain is found in many dairy products and is used mainly for its acidifying property. In Sudan, a combination of lactic acid bacteria and yeasts is responsible for the fermentation of the traditional fermented milk, *rob*. The predominant lactic acid bacteria in rob fermentation have been found to be *Lactobacillus fermentum*,*Lactobacillus acidophilus*, *Lactococcus lactis* and *Streptococcus salivarius* . The predominant yeasts have been identified as *Saccharomyces cerevisiae* and *Candida kefyr* . This association of lactic acid bacteria and yeasts is also found in the fermented milk of Ethiopia, and in the kefir and koumis milks of Eastern European countries. The taste of these fermented milks displays an alcoholic flavour and is less sour, properties that are not well appreciated in Central and East Africa. *Kivuguto* also resembles to yogurt, the well-known example of a fermented milk product.

However, the taste is different and the strains of fermenting bacteria used in yogurt are also different: *Lactobacillus bulgaricus* and *Streptococcus thermophilus*. Taste is the sole reason why people of the Great Lakes Region (Africa) continue to prefer the traditional *kivuguto*, although yogurt is also sold at the local market. Of all the fermented milks discovered to date, the one most similar to *kivuguto* is the *filmjölk* of Sweden, in terms of both consistency and flavour. When people from Central Africa stay in Europe, they usually consume this milk more than they do yogurt. *Filmjölk* is a mesophilic fermented milk, which, like *kivuguto*, is also composed of two strains of *Lactococcus lactis* and*Leuconostoc mesenteroides*. *Filmjölk* flavour is also very similar to *lben* of Morocco, a buttermilk fermented by *Lactococcus lactis*subsp. *lactis* biovar *diacetylactis*, *Leuconostoc lactis*, *Leuconostoc mesenteroides* subsp. *cremoris* and *Leuconostoc mesenteroides* subsp. *dextranicum*.

Acidification and physico-chemical properties of kivuguto

The acidification process of *kivuguto* is created by two strains: CWBI-B1466 *Lactococcus lactis* and CWBI-B1470 *Leuconostoc pseudomesenteroides*. The third strain responsible for the acidification of *kivuguto* – *Leuconostoc mesenteroides* subsp. *mesenteroides* – does not grow on its own in milk. In order to reach the isoelectric point at pH 4.6, CWBI-B1466 *Lactococcus lactis* takes 8 h whilst CWBI-B1470 *Leuconostoc pseudomesenteroides* takes 14 h. At that point, the titratable acidity is more than 80 °D. After 4 h of fermentation, a coagulum begins to form. For a period of up to 8 h, the *kivuguto* appears as a semi-solid pleasantly sour product. However, before it is suitable for drinking, it needs to be blended. *Kivuguto* keeps well for up to 36 days at 4 °C, maintaining its good flavour; its visco-elastic properties also increase with the duration of storage.

THE DAIRY INDUSTRY

Dairy policy

The livestock (animal resources) sub-sector falls under the responsibility of the Ministry of Agriculture and Animal Resources (MINAGRI). Since, 2006, the activities of this sub-sector have been implemented by the Rwanda Animal Resources Development Authority (RARDA), as part of the global programme of the National Animal Resources Policy. In Rwanda, there is no dairy sub-sector as such, but cattle farming plays a very important role in the socio-economic activities of the country. Farming of goats, sheep, pigs, chickens, fish and rabbits is less well developed. To date, the livestock (animal resources) sub-sector has been represented mostly by cattle development projects. In 2003, the Dairy Cattle Development Support Project (PADEBL) was created as a separate project with the specific objective of satisfying domestic demand for milk and beef, as well as of increasing the income of farmers.

This project contributed to the implementation of the "One cow per poor family" (Girinka) programme, launched in 2006 to provide cows for poor families. PADEBL is also responsible for milk quality management: it created and now owns many milk collection centers (MCCs), each with its own cooling system. PADEBL is financed by the African Development Bank (ADB). From 2010, MINAGRI restructured its activities. MINAGRI comprises a Permanent Secretariat to coordinate the ministry's policies and programmes through two units at central level: the Strategic Planning & Programme Coordination Secretariat (PPS) and the Internal Resource Management & Finance Unit; and through two boards at decentralised level: the Rwanda Agriculture Board (RAB), and the National Agriculture Export Development Board (NAEB). RAB comprises four departments: Agriculture Extension (formerly RADA – the Rwanda Agriculture Development Authority), Livestock (Animal Resources) Extension (formerly RARDA), Research (formerly ISAR) and Infrastructure & Mechanisation. Within the new structure, the sub-sector coordinates the issuing of certificates authorising imports of domestic animals, semen, fertilised eggs, seeds, plants and cuttings and other animal husbandry products. It also works in livestock research, training and information.

In addition to inheriting the "One cow per poor family" (Girinka) programme, the sub-sector is also responsible for the "One cup of milk per child" programme, and the "Poultry and Hatchery, Insemination and Small Stock" programme. The Girinka programme aims at enabling every poor household to own and manage a productive dairy cow, enabling the family to improve their livelihood through increased milk and meat production and to improve the soil fertility of their land for their crops using the available manure. The "One cup of milk per child" programme is a school milk programme funded by the European Union (EU) to provide milk to children in nursery and primary schools in order to address the country's problem of severe child malnutrition.

Milk production and processing

Milk production in Rwanda was estimated to have reached 372,619 tons by 2009 (DMS, 2009), following tremendous efforts made by the government and other stakeholders. In order to promote milk production in the country, emphasis was placed on bovine genetic improvement (strengthening the use of artificial insemination, and the importing of exotic cows). Most of the milk in Rwanda (48%) is produced in traditional or extensive grazing system in the Eastern Province.

The 2006 Agriculture Survey indicated that there were 1.2 million cattle in Rwanda, of which 13.6% were milking cows. The survey also indicated that there were a total of 157,479 milking cattle; improved breeds constituted 23% of that group, pure breeds 6% and crosses 17%. In 2010, the NISR (National Institute of Statistics of Rwanda) revealed the number of dairy cattle in the country to be 207,507, corresponding to an increase of 25%. The overall total number of cattle was confirmed to be 1,548,521, with 15.5% of these being represented by exotic breeds (Modderman, 2010). The Rwandese dairy industry currently consists of five dairy companies: Inyange Industries, Nyanza Dairy, Eastern Savannah, Rubilizi Dairy, Masaka Farms plus several cheese processing units mainly located in the Western Province (20 cheese processing units), with a few others being found in other parts of the country (RBS, 2011). Two of these cheese processing units are located in Nyanza District and two more are found in Gasabo District. According to Rutamu (2008), the milk processing capacity available in Rwanda is estimated at 188,000 l per day, but only 3.3% of this capacity is used. Today, these data may be revised, if we take into consideration milk processed for cheese making and the existence of new dairy processing units. Milk technology is applied in the Rwandese dairy industry in the processing of the following dairy products: pasteurized milk, skimmed milk, cream, flavored milk, fermented milk and yogurt, UHT milk, and cheeses such as Gouda, butter, ice cream.

Milk consumption and importation

High population growth in Rwanda has led to an increase in demand for livestock products that is higher than can be satisfied by domestic production. This has forced the country to invest its few financial resources in the massive importation of milk. From 1999 to 2007, the trend in milk imports decreased from 1,280 t to 450 t . It is very difficult to make an accurate estimation of milk consumption in Rwanda, because a large volume of milk is sold in the informal market. In 2007, approximately 62,000,000 l were consumed on farms and about 35% of raw milk was lost due to spoiling before reaching the market or being processed . In 2008, a chain analysis estimated that 96% of all dairy products are sold in the informal market, *e.g.* in small informal milk shops. Kigali city has more than 1,500 milk shops (Modderman, 2010). Milk is consumed either fresh or after coagulation. As Rwandese people prefer fermented milk, it is very difficult to estimate the percentage of fresh milk consumed out of the 96% of all dairy products sold at the informal market.

The dairy value chain

Michael Porter was the first to use the concept of the "value chain". Within a firm or organisation, both the service provided and the product made add value according to the activities and processes employed by the organisation.

Value chains traditionally refer to the chain of activities that products pass through as they gain value before reaching the consumer . Along the chain, the organisation is split into primary activities and support activities. The dairy value chain in Rwanda follows two main segments of actors, with the segments being composed of multiple nodes. The two segments are linked to input providers organised by the Ministry for Agriculture and Animal Resources (MINAGRI) through the Rwanda Livestock sub-sector and other stakeholders (other state organisations, NGOs and independent traders).

Dairy value chain input providers

Since, 2006, under the Ministry of Agriculture, RARDA (Rwanda Animal Resources Development Authority) and ISAR (Institute of Agronomic Sciences of Rwanda) have supported the dairy value chain at different levels. The mission of RARDA was to contribute towards the growth of animal production through the development of appropriate technologies, and by providing advisory, outreach and extension services to stakeholders in the animal resources sector. Since, 2003, PADEBL (Dairy Cattle Development Support Project) has contributed to a sustainable increase in milk and meat production, through the introduction of appropriate technological packages into stockbreeding in villages and peri-urban areas.

ISAR was the body responsible for research and development at MINAGRI. ISAR's mandate was to carry out research on agronomy, and on animal and environmental sciences (ISAR, 2010). The Rwanda Bureau of Standards (RBS), under the Ministry of Trade and Industry (MINICOM), is the statutory body charged with the enforcement of standards and the certification of all products and services in the country (RBS, 2011). The Rwanda Cooperative Agencies (RCA) is a public institution in charge of the promotion, registration and regulation of cooperatives in the country. The livestock sub-sector receives a lot of support from RCA for the dairy development, since, many cooperatives are involved in the dairy chain. Other stakeholders in the livestock sub-sector are ADB (African Development Bank), BNR (Banque Nationale du Rwanda), BPR (Banques Populaires du Rwanda), BRD (Rwanda Development Bank) and other commercial banks; NISR (National Institute of Statistics of Rwanda); independent traders (veterinarians, transporters, etc.); farmers' cooperatives and many NGOs (non-governmental organisations).

Dairy value chain actors. Informal actors are present throughout the two segments. From the farmer to the consumer, dairy activities are conducted by dairy processors, supermarkets, collection centers, transporters, and traders and retailers. Farmers are categorised according to the feeding system they use: free-grazing, semi-grazing or zero-grazing. Individual cattle farmers generally practice free-grazing and they produce milk for domestic consumption. Sometimes, surplus milk is sold to a retailer, a transporter or a cooling system owner. This cattle husbandry system represents the vestige of the old system; MINAGRI discourages this system and supports instead zero grazing. Zero grazing is practiced mostly by peri-urban farmers. These farms are well equipped around Kigali city, and the owners have invested both in large numbers of cattle and in equipment in order to supply dairy processing units. Semi-grazing systems have sufficient land and, with a low level of investment, they are able to produce milk to supply a retailer, a transporter, a cooling system owner or a processor.

The processor then supplies dairy products to supermarkets. Semi-grazing and free-grazing husbandry farmers are categorised as "small scale farmers" along the dairy value chain. Cooling systems mainly belong to PADEBL and to some farmers' cooperatives. These cooling systems are held in milk collection centers (MCCs). Transporters are mainly traders, who use either bicycles or vehicles to transport milk. However, some transporters are paid by farmers only for collecting milk from the farm and then taking it to the collection center or to a retailer.

These transporters supply milk from small farmers to milk shops or directly to consumers. Those with vehicles sell milk to processors, direct from the collection centers. Milk retailers consist of bicycle milk traders selling milk from door to door, milk shops selling only milk or tea, and ordinary shops selling milk alongside other household products. TechnoServe (2008) estimated the profit margin within each node of the dairy value chain to be as follows: 15% for input providers, 62%-28%-44% for dairy farmers depending on how cows are fed respectively open- semi- zero- grazing, 15-25% for transporters, 6% for cooling system owners, 16% for processors, 10% for sellers of raw milk and 15-20% for supermarkets. To date, the dairy value chain in Rwanda has been scattered; nevertheless an organised structure is in the process of being established. PADEBL began a reorganisation of the dairy industry in 2009 by setting up some milk collection centers equipped with cooling systems. However, thus far, the number of MCCs equipped in this way has remained below the level needed.

MILK DEVELOPMENT STRATEGIES

The Rwandese government development programme, Vision 2020, aims to transform agriculture from a subsistence level to a productive, high-value, and market-oriented farming industry, which is environmentally friendly and has an impact on other sectors of the economy. In this regard, a new value chain will take into account the quality of milk along the value chain. Vision 2020 takes into account the growing pace of the Rwandese economy in general as well as the Livestock sub-sector, focusing on the target of exporting dairy products. Consequently, the Livestock sub-sector will set up a developmental programme. It will also accelerate the restructuring of the dairy industry in the whole country with regard to the cold chain, as milk is a very temperature sensitive product.

At the same time, distributing and trading milk in ordinary plastic material and on bicycles will be prohibited. The establishing of MCCs in all the dairy basins of the country is seen as the best solution. Mobile cold storage system respecting HACCP standards may replace the current set of standards for milk transport. Establishing cold-storage systems on farms will provide another solution in the fight against microorganisms that pose a threat to human health. *Escherichia coli*, *Listeria monocytogenes*, and *Salmonella*are destroyed by pasteurization or by simple boiling. However, microorganisms such as *Staphylococcus aureus*, despite being killed by pasteurization, are still harmful due to the toxins they produce.

Thus, the protection system close to the farm must be based on a strategy of prevention. RBS must assist with the implementation of standards from the cow on the farm to the consumer. All these measures will be made possible by improving firstly, cattle breeding, genetics and nutrition and secondly, standards of milk preservation, storage, packaging and transportation. For that purpose, a business plan for the period 2008 to 2020 has been established as part of a dairy developmental programme introduced by the Livestock sub-sector, based mainly on the identified dairy basins. The new structure of the Livestock sub-sector has served to highlight the targets fixed from 2008 to 2020 in this dairy development programme, aimed at stimulating the development of a modern livestock industry in Rwanda through sustainable growth, increasing value added, and access to markets. Five pillars have been identified for the support of this programme :

- Community livestock infrastructure support;
- Public livestock infrastructure support;
- Support for the "One cow per poor family" programme;
- Support for productivity enhancing technologies;
- Capacity building.

The programme has planned certain objectives, with Rwanda being exporters of milk and milk by-products by 2020. The programme will set up further milk collection centers, livestock watering systems and markets, veterinary clinics and quarantine stations in all dairy basins. Poor families will benefit from having cross-breed heifers to support food security and income generation.

Training in genetic and feeding techniques will be provided for staff and farmers in the livestock industry, in the form of study tours, short courses, seminars, conferences and workshops. The ultimate goal is the integrated development of the dairy industry (DMS, 2009) in which all stakeholders must play a role in coordinating the production, processing and distribution of high quality milk and its by-products.

Little research has been conducted on *kivuguto*. In the present paper, the traditional technology used to make this fermented milk as well as its by-products has been described. The dairy industry in Rwanda has also been discussed. Proposed improvements in the development of milk production and the dairy sub-sector in Rwanda at all levels and with all stakeholders have been outlined. Information provided on the value of the whole dairy chain raises some interesting issues in terms of trade in Rwanda and in the region. This study is part of a global research project on *kivuguto* milk, with the aim of producing this foodstuff in industry.

UNDERSTANDING FUNCTIONAL FERMENTED MILK BEVERAGES

Milk and milk-derived products have constituted a significant part of the diet of all ethnic groups at all ages, including a proportion of the food provided to infants. Amongst those milks products, fermented milks are of great importance worldwide because of their nutritional, organoleptic and shelf-life properties that are significantly improved when compared with its raw material i.e. milk.

Originally fermented milks were developed as a means of preserving nutrients of milk. Fermented milks including dahi, yogurt are considered as an ideal vehicle for the delivery of many beneficial microorganisms viz probiotics and prebiotics in addition to the microflora of human gastrointestinal tract, therefore fermented milk is the most popular group of functional food. Fermented milk products contain various nutritional components such as bioactive peptides, antioxidants, vitamins, specific proteins, oligosaccharides, organic acids, highly absorbable calcium, conjugated linoleic acid and other biologically active components with an array of bioactive functions: modulating digestive and gastrointestinal functions, haemodynamics, controlling probiotic microbial growth and immunoregulation.

Many changes occur to the components of milk during fermentation, although theoretically there should be no significant difference between the gross composition of unfermented and fermented milk, it has been proved that fermented milks have a higher nutritional composition compared to regular milks, as a result of bacterial fermentation (mainly lactic acid bacteria) under controlled conditions.

Furthermore fermented milks have a large variety of flavour and resulting from lactic acid production and the presence of specific aromatic components. Acetaldehyde, diacetyl and acetone give particular flavour to the fermented milks. The nutritional and health benefits of fermented milk products are the result of biologically active components that are present in native milk and also, due to their suitably modulated activities produced by the action of lactic acid bacteria. According to Prajapati et al. modification of nutritional value is related to changes made by fermentation as well as changes made by processing parameters. Consumption of fermented milks has increased significantly around the world and nowadays various popular ingredients of functional significance are being incorporated into cultured dairy products to enhance their market value. Since, consumption of functional foods containing nutraceuticals is being highly encouraged, thus fermented milks produced with incorporation of these ingredients with specific health benefits are of potential interest. Recently, a lot of research work is being carried out around the globe regarding the effect of various added ingredients like fruits, inulin on fermented milks. Also, it is imperative to know a meaningful dose-benefit relationship associated with a specific fortified food.

Considerable progress has been made in demonstrating certain beneficial effects of fermented milk in animals, probably due to the nutritional changes occurring in milk during fermentation. However, unequivocal experimental or epidemiological evidence still needs to be gathered to substantiate claims of similar effects in humans.

Fermented milk products have been reported to have therapeutic properties like anti-cholesterolemic, anti-carcinogenic and anti-cariogenic properties beyond their basic nutritive value. They, contributing to a variety in our gustative desire, have been recognised to provide important nutrients and considered superior over non-fermented dairy products in terms of nutritional attributes as the microflora present produce simple compounds like lactic acid, amino acids and free fatty acids that are easily assimilable. Originally fermented milks were developed as a means of Lactic Acid Bacteria (LAB) and they have been widely used as starter culture for the manufacturing of various fermented dairy products such as dahi, lassi and whey beverages. LAB and their food products are thought to confer a variety of important nutritional and therapeutic benefits and have many documented health promoting or probiotic effects in humans such as inhibition of pathogenic organism, antimutagenic and reduction of blood cholesterol. Those LAB with scientifically supported health claims are defined as probiotic and have an increasingly high market potential in preserving nutrients.

FERMENTED MILK PRODUCTS

Fermented milks are manufactured throughout the world and approximately 400 generic names are applied to traditional and industrialised products but in actual essence the list may only include a few varieties.

Lactic fermentations that include -

(a) Mesophilic type, e.g., cultured buttermilk, filmjolk, tatmjolk and langofil;

(b) Thermophilic type, e.g., yoghurt, Bulgarian buttermilk, zabadi, dahi and

(c) Therapeutic or probiotic type, e.g., acidophilus milk, Yakult, ABT, Onka, Vifit; products within this group constitute by far the largest number known worldwide;
   (a) Yeast – lactic fermentations (kefir, koumiss, acidophilus yeast milk); and
   (b) Mould – lactic fermentations (villi).

Certain closely related products are manufactured from fermented milks by de-wheying; examples include labneh, skyr, ymer and shrikhand. According to a study by global market analyst Euromonitor International, global sales of dairy products reached □211.5 billion. The manufacture of cultured dairy products represents the second most important fermentation industry (after the production of alcoholic drinks). A dynamic category, fermented dairy drinks were reported to grow at six times the rate of total dairy growth between 1998 and 2003 in value terms. Also, probiotic drinking yoghurt was the fastest growing dairy product sector between 1998 and 2003, followed by soy milk, (spoonable) probiotic yoghurt, flavoured milk drinks with juice and fermented dairy drinks.

The increasing demand from consumers for dairy products with 'functional' properties is a key factor driving value sales growth in developed markets. This led to the promotion of added-value products such as probiotic and other functional yoghurts, reduced-fat and enriched milk products and fermented dairy drinks and organic cheese.

There are several principal reasons for the success of fermented dairy products, which relate to nutrition and health, versatility and marketing. The consumption of milk drinks and fermented products has been recently reviewed by the International Dairy Federation: It is quite clear from the data that the consumption of fermented milks has generally increased around the globe over a period from 2001 to 2004.

The occurrence of various bioactive peptides in fermented milks, e.g., yoghurt, sour milk and "Dahi", has been reported in many studies. ACE-inhibitory, immunomodulatory and opioid peptides, e.g., have been found in yoghurt and in milk fermented with a probiotic L. rhamnosus strain. Also, ACE-inhibitory peptides have been detected in yoghurt made from ovine milk and in kefir made from caprine milk.

Whey-based beverages

Whey is a byproduct from cheese and casein production. It is an important source of lactose, calcium, milk proteins and soluble vitamins, which make this product to be considered as a functional food and a source of valuable nutrients. Whey products have certain essential amino acids, good digestibility, and protein efficiency index higher than 3.0. Vitamins such as thiamin, riboflavin, pantothenic acid, vitamin B6 and B12 are also present. Functional properties of whey proteins, such as emulsifying, water/fat holding, foaming, thickening and gelling properties, also make them interesting to be used as a food ingredient. Due to their functional properties, whey solids/ whey as such could be used in conjunction with fermented milks.

Several studies have focussed on the use of milk whey in yoghurt making and use of whey powder or whey–milk powder mixtures. This process leads to the increase of milk total solid content in order to provide better consistency, texture and creaminess to the product. In other studies, replacement of skimmed milk by whey protein concentrates (WPC) and milk protein concentrates (MPC) were studied.

Thus, yoghurts with different mineral and protein composition were obtained. It was observed that these components are of decisive importance in the fermentation and gelling process and also in the type of gel obtained. However, yogurt microorganisms should be plenty and alive in the final product. In the present scenario of consumption of fermented whey drinks such as Molke in West Germany, and Rivella in Switzerland and these products are showing increasing trends in most of the countries around the world. Keeping in view increased demand for soft drinks and juices these days in India, there is tremendous scope and need to exploit commercial production of these fermented whey beverages since, it is the best proposition to convert largest by-product (whey) of dairy industries into value-added product by simple and indigenous processes.

Whey beverages and drinks

A variety of whey beverages such as plain, carbonated, alcoholic, soya and fruit types have been successfully developed and marketed all over the world, because they hold great potential for utilising whey solids - advantages of fermenting whey into various types of beverages have been highlighted by Gandhi (1996).

In India, a number of refreshing whey drinks and beverages have been developed that include whevit, acido-whey, whey-based fruit beverages, whey-based soups, whey-based lassi and whey-based sport beverage. These beverages are preferably prepared from paneer/chhana whey, which is acidic and has low protein content.

Whevit

Whevit, an orange, pineapple, lime or mango flavoured alcoholic drink from whey, was developed at National Dairy Research Institute (NDRI), Karnal. For its manufacture, fresh whey is efficiently separated in cream separator, deproteinised by steaming and cooled to room temperature. To the deproteinised and clarified whey, 22-23% of 50% sugar solution is added followed by 2-2.1% of 10% citric acid, colour and flavour. It is then fermented by incubation at 220C for 14-16 h with a 1% culture of Saccharomyces cerevisae. The product is bottled, pasteurised, cooled and stored at low temperature (5-10oC). The final product contains 0.5 to 1% alcohol.

Acido-whey

For the manufacture of acido-whey that is a non-alcoholic whey drink, deproteinised whey is fermented with a culture of Lactobacillus acidophilus and Lactobacillus bulgaricus. Sugar and flavour are then added and the product is heated at 75oC for 5 min, cooled to 5oC, packed in pouches and stored in refrigerated conditions.

Whey-based fruit beverages

Whey-based fruit beverages are manufactured by mixing of appropriate fruit pulp/juice or juice concentrate and processed whey followed by proper heat processing and packaging . The more sophisticated technological approaches include the painstaking selection of proper flavour or flavour blends, complex whey treatment operations and often the inclusion of additional nutrients to increase the consumer appeal.

The sedimentation problem can be countered by centrifugal clarification to remove all sedimentable, fine curd particles. The pH of the final product should be controlled accurately, especially if subsequent severe heat processing is to be used.

To minimise heat induced whey protein precipitation, adjustment below pH 3.8-3.6 is necessary as the whey protein fraction becomes resistant to coagulation below this critical range.

Whey-based soups

The common sequence of the operations in thc production of a whcy-based vegetable soup is: blending of the vegetable and corn flour in whey followed by heat processing.

The time-temperature combination for cooking of vegetables, corn flour and seasoning is important for proper dispersion of vegetables, gelatinisation of starch and flavour perception of soups. The development of long shelf life soup involves proper sterilisation of soup. Whey based soup powders can be manufactured by cooking of vegetables in concentrated whey, mixing in it fried seasonings and gelatinised starch followed by spray drying. However, commercial soup powders available in the market comprise blending of dried vegetables in gelatinised starch.

Soups are served as appetisers before meals as they stimulate the secretion of gastric enzymes that leads to feeling of hunger. In market a large number of ready-to-make soup mixes are available to suit the palate of consumers. But certain additives in such soups mixes are considered harmful particularly to children. Moreover apparently they do not seem to provide quality nutrients and utilisation of whey for soup preparation is an attractive possibility. The process for the manufacture of whey-based soup involves blending of vegetables in whey and cooking of corn flour followed by heating. The time-temperature combination of cooking of vegetables, corn flour and seasoning is important for dispersion of vegetables, gelatinisation of starch and flavour perception of soup respectively .

The developed product could be stored for a week under refrigeration and UHT treatment can be adopted to improve the shelf-stability. Paneer and cheese whey were utilised for the potato-carrot-tomato and spinach soups. Cheese whey was preferred for the manufacture of vegetable soups than paneer whey . The reason could be the low pH of paneer whey that resulted in acidic product not usually compatible with most vegetables. Whey- based soups have been reported to be more viscous as compared to water-based most probably gelation of whey proteins on heating. Whey-based soups require less amount of salt, thickener and fat.

Technology for manufacture of retort processed low fat tomato-whey soup has been developed recently at the NDRI. Alam et al (2002) reported the technological aspects for the manufacture of tomato whey soup using paneer whey. Few years back Amul has introduced UHT processed tomato-whey soup in Tetra Pak and last year VITA has launched tomato-whey soup in polystyrene cup in Haryana.

Whey-based lassi

Whey-based lassi, in which up to 60% of milk is replaced with whey, has also been developed at NDRI. The product formulation requires addition of pectin, CMC and trisodium citrate. The product has 2.0% fat, 1.8% protein, 4.6% lactose and about 23.0% total solids. This product can also be UHT-processed for long shelf life.

Probiotic whey beverages

A probiotic whey beverage (with addition of sugar and pectin) have been produced using probiotic cultures Lactobacillus reuteri and Bifidobacterium bifidum. The beverage fermented by probiotic strain Bb-12 obtained lower sensory score than beverages fermented by strains La-5 and Lc-1. Lactobaciluus acidophilus, YG culture, Lactobacillus rhamnosus and Bifidobacterium animalis subsp. lactis were used to produce a fermented whey product. A fermented whey drink was prepared by using yoghurt cultures and co-cultures Streptococcus thermophilus, Bifidobacterium animalis subsp. lactis. Streptococcus thermophillus and Lactobacillus delbrueckii sbsp. bulgaricus, has a high potential culture for whey fermentation. A cultured whey was prepared from raw milk whey using B. bifidum having self life of 7 days at refrigeration temperature. Whey-based probiotic product was developed by using L.reutri and B. bifidum with a shelf life of 14 days.

Benefits of whey-based fermented drinks

Whey is an excellent growth medium for Lactic Acid Bacteria to ferment lactose in whey to form lactic acid.

Whey is a genuine thirst quencher unlike most soft drinks.

Whey as a drink can replace much of the lost organic and inorganic salts to the extracellular fluid.

Whey is rapidly adsorbed due to absence of fat emulsion.

Whey has been used to treat various ailments such as arthritis, liver complaints and dyspepsia.

It also possesses almost all the electrolytes of Oral Rehydration Solution (ORS), which is invariably used to control dehydration.

On fermentation with LAB, it becomes a suitable drink for lactose-intolerant people.

Fermentation of whey with LAB also masks the effect of curdy flavour of whey.

At industrial scale, large volumes of whey can be used directly from paneer/ cheese vats, thus eliminating transportation and disposal problems.

Conversion of whey into beverages involves very simple processes.

Utilisation of whey generates additional revenue to the dairy plant.

Above all, its utilisation also solves the problems of environmental pollution.

Functional dairy beverages

A probiotic is defined as a 'living organism which when administered in certain numbers exerts health benefits in the host' (FAO, 2001). Owing to this property, incorporation of probiotic micro-organisms in dairy foods has increased rapidly during the last two decades.

Consumption of probiotic bacteria via food products is an ideal way to reestablish the balance of intestinal microbiota. These include alleviation of lactose intolerance symptoms, lowering cholesterol, curing antibotic-associated diarrhoea, prevention of intestinal tract infections, prevention of colon cancer, control of rotavirus, prevention of ulcers related to Helicobacter pylori, improvement of immune system, irritable bowel syndrome and antihypertensive effects.

In order to produce therapeutic benefits, a suggested range for the minimum level for probiotic bacteria in probiotic milk is from 106 to 107 colony-forming units . In recent years, probiotic beverages based on fruit juice, cereal products and daily dose dairy drinks have also become popular commercially. Today, a wide range of probiotic products is available for consumers in the market.

Non-probiotic dairy beverages with added bioactive components

a-linoleic acid (C18:3 n-3, ALA), eicosapentaenoic acid (C20:5 n-3, EPA) and docosahexanoic acid (C22:6 n-3, DHA) are the principal members of the Omega-3 fatty acids. Europe's first commercial Omega-3 enriched milk was introduced by Dawn Dairy (Ireland) and the desired amount of Omega-3 was added at the final stage of bottling by means of a dosing system developed by Tetra Pak. Parmalat, an Italian dairy company, launched Omega-3 enriched milk 'Omega-3 Plus' (containing 80 mg Omega-3 per litre of milk) in 1998. Consumption of Omega-3 enriched semi-skimmed milk at a dose of 500 mL per day.

Fortification of Omega-3 enriched milk with vitamins A, C and E is a common application. In order to attract consumers, chocolate flavoured milk fortified with Omega-3 has been introduced in the functional foods market by Neilson Dairy (Dairy Oh, Canada) and Parmalat (Beatrice, Canada). Two potent ACE-inhibitory peptides, Valine-Proline-Proline (VPP) and Isoleucine-Proline-Proline (IPP), derived from caseins during milk fermentation with Lactobacillus helveticus and Saccharomyces cerevisiae, are responsible for the anti-hypertensive activity shown by Calpis sour milk (Calpis Co. Ltd, Tokyo, Japan). Other examples of commercial dairy-based beverages with added bioactive peptides are Evolus. The former product is manufactured by incorporating two tripeptides, Val-Pro-Pro and Ile-Pro-Pro, and is claimed to reduce blood pressure upon regular consumption. The latter product contains the same tripeptides added to Evolus plus plant sterols, which help to reduce blood cholesterol levels.

Melatonin is a naturally occurring hormone found in animals and in some other living organisms, including algae. Melatonin controls the body's day and night rhythm. A Finnish company, Ingman Dairy, introduced the world's first ever high-melatonin premium milk (under the Night Time brand) in 1999. The same company produces high-melatonin milk based on the standardised milking system at night. A similar product has been launched in Japan, and the UK has two milk brands inspired by the Finnish example.

Conjugated linoleic acid is found almost exclusively in animal products, with a natural level of approximately 6 mg / gm of fat. Normal daily intake of CLA in the diet is 150–400 mg / gm addition of linoleic acid at a level of 0.1% increased cis9-trans11-CLA content of non-fat yogurt significantly without affecting the sensory properties of the final product.

Other fermented milk beverages Fruit lassi

Cultured dairy products are an excellent medium to generate an array of products that fit into the current consumer demand for health-driven foods. Owing to expanding market share and size of dairy companies, there has been a reduction of clearly structured markets i.e. merging of dairy products and fruit beverage markets with introduction of juiceceuticals that include hybrid products like fruit-based cultured milk beverages. In India lassi made out of dahi is a widely consumed fermented milk beverage, popular in all parts of the country and has a great potential in the domestic as well as overseas markets. However, problems like short shelf life, post-acidification, whey syneresis, hinder the market saleability of lassi. Accordingly, studies were conducted for the production of fruit lassi with extended shelf life using biopreservatives. Khurana (2006) developed a process for making fruit lassi, using buffalo milk. Standardised buffalo milk was heated to 90°C for 10 minutes followed by cooling to 30°C and inoculated with 1.5 per cent culture and incubated at 300C for 10-12 hrs.

The curd obtained was then mixed with sugar syrup and fruit pulp. Homogenisation was done followed by packaging, cooling and storage at 4°C. For further stabilisation and improvement of consistency of each type of fruit lassi exopolysaccharide producing (EPS+) cultures were used along with pectin at different levels.

An enhancement in the rheological and overall sensory characteristics of all the 3 types (banana, mango and pineapple) of fruit lassi was observed with increase in proportion of EPS+ culture up to a certain level. Nisin and MicroGARD were used to extend the shelf life of fruit lassi up to 30 and 50 days, respectively as compared to control mango lassi which had a shelf life of 15 days at 4±1°C.

The technology developed for manufacture of fruit lassi with extended shelf life appeared to have considerable potential to facilitate commercial manufacture and marketing of this popular fermented milk beverage. Inclusion of fruits and artificial sweeteners in lassi would not only enhance nutrition, help diabetic/ obese people, aid product diversification but also help in curtailing the post-harvest losses in fruits.

Such a product would not only serve as a low calorie-quick meal snack but also offer stiff competition to expensive soft drinks in the beverage market.

Cereal-based fermented dairy beverages

Cereal grains constitute a major source of dietary nutrients all over the world. Addition of cereals into milk not only enriches its mineral value but also supplements fibre.

Fermentation further enhances the nutritive value, palatability and functionality of cereals by reducing the antinutritional factors. Development of technologies for the manufacture of cereal-based fermented milk beverages will lead to the utilisation of underutilised cereals like sorghum, pearl millet, and finger millet. Raabadi, prepared by cooking cereal (maize) flour with buttermilk, is a traditional popular beverage of Haryana, Punjab and Rajasthan. The technology of producing these traditional fermented foods from cereals and milk remains a household art. Traditional process of raabadi preparation yields a product with limited shelf life (one to two days) with unpredictable sensory quality.

A technology has been developed at NDRI for manufacture of raabadi-like fermented beverage using cereals like pearl millet/sorghum/wheat possessing health attributes. Development of raabadi-like fermented milk beverages was based on the selection of milk solids source; selection of levels of milk solids and cereal solids; stage of addition of cereal solids; fermentation conditions and stabilisation of developed product in terms of preventing sedimentation; and wheying-off in the product during storage. Cereal solids were incorporated in milk at two stages i.e. before fermentation with dahi/yoghurt culture or after fermentation.

The fermented mass will be blended into beverage by the addition of water, salt and spices. The cereal-based beverage thus developed was having very good sensory qualities and it was accepted well by the consumers but the only limitation of the beverage was its limited shelf life.

Attempts were made to increase the shelf life of the product in which, this beverage was made in ready to reconstitute form. Thus, it not only increased the shelf life of the beverage but at the same time it aided to the convenience in use.

In another study to increase the shelf life of the liquid beverage, various preservatives viz. biological, chemical and thermal treatments were tried and the shelf life increased to about a month at refrigerated temperature. Gupta et al. (2007) prepared raabadi from buttermilk and mothbean by two methods. Buttermilk was mixed with mothbean flour to get homogeneous paste and this was diluted with plain water and boiled for 7 min with constant stirring while salt was added during stirring.

In the first method the mix was fermented before cooking and in the second method the mix was fermented after cooking. They reported that 4 h fermented and cooked and 12 h cooked and fermented raabadi had better acceptability. Kindumu is a fermented milk beverage popular in central African region and is prepared by sun drying the mixture of fermented milk and germinated/non-germinated sorghum flour. Grewal and Chauhan (1993) prepared Soy raabadi by blending autoclaved dehulled soybean slurry and the curd mass (obtained from reconstituted skim milk).

The blend was added with water and fermented at 30°C for 12 h. Soy raabadi appeared to be a low cost wholesome food which was free from beany flavour and rich in protein, fat and dietary essential minerals. Mugocha et al. (2000) developed a composite finger millet and milk-based fermented beverage. Various parameters including level of finger millet gruel in skim milk, type of starter culture, incubation temperature and so on were optimised to develop the finger millet-dairy beverage

PROCESS FOR THE PRODUCTION OF A FERMENTED MILK PRODUCT

FIELD OF THE INVENTION

The present invention relates to a process for manufacturing fermented milk products as well as to lactic acid bacteria capable of producing diacetyl and/or acetoin at thermophilic temperature conditions.

Background of the invention

Lactic acid bacteria are extensively used for production of fermented foods and they greatly contribute to flavour, texture and overall characteristics of these products. An old and well known example is yoghurt which probably originated from the Middle East and which still makes up more than half of the fermented milk production - or approximately 19 million tons in 2008 (source: Euromonitor). Fermented milks as e.g. yoghurts are popular due to the healthy image and pleasant sensory properties.

Initially the souring of milk was done by natural fermentation. With the advent of the knowledge of bacteriology and nutrition, fermentation process emerged as a well developed controlled sequence of changes in milk with the use of selective microorganisms. Selective organism produces lactic acid and may impart other ben eficia l effect to the prod uct. Th e use of d ifferent microorgan isms led the development of a wide range of milk products viz. dahi, yoghurt, shrikhand, lassi, kefir, koumiss, yakult, laben, etc. In India, Dahi is being produced with varieties of taste varying with region-to-region and individual food habits.

Dahi is a popular fermented milk in India, Nepal and other Asian countries It is still made in every household in villages as well as in urban areas using traditional methods. Dahi is also known as "Curd", in particular in the southern part of India.

The product is typically made from pasteurized or boiled milk, inoculated with dahi as starter left over from the previous day. Incubation is carried out at ambient temperature usually overnight.

In contrast to the household production, more controlled processes for the dahi making in the organized (industrial) sector have been developed. Fresh, sweet, good quality milk is received, pre-heated and subjected to filtration and clarification.

The milk is standardized to 2.5 to 3.0 % fat and 8.5-10.5% solids not fat, pre-heated to 60-65°C and homogenized. Homogenization is typically a two-stage process, with a first stage at 1500 psi (100 bar) and a second stage at 500 psi (35 bar) both at 60-65°C

Subsequent pasteurisation may be carried out at any suitable combination of time and temperature, for example at 85-90°C for 15-30 minutes or at 90-

95°C for 5-10 min. After cooling down to 22-25°C and the milk is inoculated with 1 -2 percent of specific dahi starter culture. It is then put in suitable packaging containers of the appropriate size and incubated at 22-25°C for 16-18 hours.

After proper setting of the dahi, the acidity of dahi reaches 0.6 to 0.7 percent and a firm curd is formed. The curd is cooled, for instance by circulating chilled water or air (a blast chiller) around the containers and then transferred to the cold room maintained at about 4-5°C, or the containers are transferred the cold room directly.

Dahi is often described as the Indian equivalent of yogurt. In reality, however, there are distinct differences between the two products. In terms of flavour, yogurt is characterised by a sharp acid tartness and the characteristic "green" acetaldehyde flavour. Dahi on the other hand has a mild acidity and diacetyl is the prominent flavour component. The latter contributes to a buttery, nut-meat like flavour.

Other flavour compounds such as acetoin, acetic acid, alcohol and carbon dioxide impart a balanced round flavour to dahi. The difference in the flavour between yogurt and dahi is caused by the different starter cultures used. For the production of dahi at the lower temperatures (e.g. 22-25°C), mesophilic lactococci and in some instances leuconostocs are used as starter flora. The basic role of starter culture is to bring about acid coagulation of milk and impart the characteristic flavour.

Presence of more at least one type of lactose fermenting microorganism in the starter culture is required for the production of diacetyl flavour in dahi. A lactic culture comprising of Lactococcus lactis subsp. lactis, Lactococcus lactis subsp. cremoris and Lactococcus lactis subsp. lactis biovar. d i acetyl a ctis may be used for dahi production.

Recent developments in the industrial production of dahi have led to production processes at higher temperatures (i.e. from ambient temperature to temperatures between 37 and 45°C) in combination with shorter incubation times (from overnight fermentation to 4-8 hours). Under these conditions thermophilic starters are used (e.g. Streptococcus thermophilus and Lactococcus delbrueckii subsp. bulgaricus) leads to the formation of acetaldehyde as in yogurt.

For dahi this is not desirable. The known diacetyl producing mesophilic leuconostoc and Lactococcus lactis subsp. lactis biovar. diacetylactis strains do not grow sufficiently at the higher process temperatures and therefore the industrial dahi produced at high temperature and short incubation time contain hardly if any of the typical "dahi" flavour in terms of diacetyl and/or acetoin. It is an object of the present invention to provide diacetyl and/or acetoin producing lactic acid bacteria which are sufficiently active at the conditions of the industrial dahi process and also to provide a process for the production of dahi with the typical dahi flavour in an industrial process at higher temperatures and shorter times.

Detailed description of the Invention

In a first aspect the invention provides a process for making a fermented milk product comprising the following steps: (a) adding one or more thermophilic lactic acid bacterial strains to milk for the acidification of the milk; and (b) adding one or more lactic acid bacterial strains for the production of diacetyl and/or acetoin that it is capable of producing at least 100 mg/l acetoin and/or at least 6 mg/l diacetyl in Test A; and (c) carrying out the fermentation process in step (a) and in step (b) at a temperature range of 37 to 45°C during a period of 8 hours or less. In a preferred embodiment, the fermented milk product is dahi or curd . Dahi and curd and other fermented milk products have been described in the background of the invention. The temperature range of the fermentation process in step (a) and in step (b) is between 37 to 45°C. These higher temperatures, in particular for step (b), are preferred because this allows the process of the invention to be carried out for shorter times which is an advantage for the dairy industry. Preferably the temperature range is of 38 to 45°C, more preferably the temperature range is of 39 to 45°C, more preferably the temperature range is of 40 to 45°C. The fermentation process of the invention comprising step (a) and step (b) is carried out during a period of 8 hours or less.

The time should be sufficient to obtain the required acidification as expressed by the pH value of the milk after fermentation in step (a) and for the production of diacetyl and/or acetoin in step (b). For example, the fermentation process comprising step (a) and step (b) may be carried out during a period from 2-8 hours, preferably 3-8 hours, more preferably 3-7 hours, more preferably 4-8 hours, more preferably 4-7 hours.

The skilled person is very well capable of determining the optimal temperature and time, in combination with the dosage of the one or more thermophilic lactic acid bacteria strains in step (a) and the one or more lactic acid bacterial strains for the production of diacetyl and/or acetoin in step (b) in order to obtain sufficient acidification as well as sufficient formation of the flavours such as acetoin and diacetyl.

For the acidification of the milk under the conditions of the process of the invention, regular thermophilic lactic acid bacteria may be used. These bacteria are well known in the art and commercially available. Preferably, thermophilic lactic acid bacterial strains are selected from the group consisting of Streptococcus thermophilus and Lactococcus delbrueckii subsp. bulgaricus. The dosages of said thermophilic lactic acid bacteria are known in the art.

For the production of the flavours acetoin and diacetyl in the fermented milk product, the process of the invention comprises adding one or more of the lactic acid bacterial strains in step (b) characterised in that the lactic acid bacterial strain is capable of producing at least 100 mg/l acetoin and/or at least 6 mg/l diacetyl in Test A, i.e. at a temperature of 40°C which is the temperature of Test A. Preferably the acetoin is produced at least at 150 mg/l, more preferably at least at 200 mg/l, more preferably at least at 250 mg/l, more preferably at least at 300 mg/l more preferably at least at 350 mg/l. Preferably, the diacetyl is produced at least at 9 mg/l, more preferably at least at 12 mg/l, more preferably at least at 15 mg/l and more preferably at least at 18 mg/l. These diacetyl values are in addition to the diacetyl that may be present in the milk used in Test A. Test A has been described in the MATERIALS and METHODS section.

The lactic acid bacterial strain capable of producing at least 100 mg/l acetoin and/or at least 6 mg/l diacetyl in Test A and which is added in step (b) of the process of the invention, may belong to the genera Lactococcus and is preferably a Lactococcus lactis subsp. lactis biovar diacetylactis. The advantage of these lactic acid bacterial strains is their capability of producing flavours at the temperature of the process of the invention, i.e. between 37 and 45°C; as measured in Test A. Most preferred are the following strains: Lactococcus lactis subsp. lactis biovar diacetylactis CBS 130345, Lactococcus lactis subsp. lactis biovar diacetylactis CBS 130346 or Lactococcus lactis subsp. lactis biovar diacetylactis CBS 130347.

It was surprisingly found that these strains were able to produce flavours such as acetoin and/or diacetyl at the temperature of the process of the invention, i.e. between 37 and 45°C. The dosages of said one or more of the lactic acid bacterial strain added in step (b) of the process of the invention may be determined by the skilled person and will depend also on the fermentation temperature and time of the process of the invention.

The fermented milk product that may be produced by the process of the invention, preferable dahi or curd, comprises preferably at least 100 mg/l acetoin and/or at least 6 mg/l diacetyl. Preferably the acetoin is at least 150 mg/l, more preferably at least 200 mg/l, more preferably at least 250 mg/l, more preferably at least 300 mg/l more preferably at least 350 mg/l. Preferably, the diacetyl is at least 9 mg/l, more preferably at least 12 mg/l, more preferably at least 15 mg/l and more preferably at least 18 mg/l.

In a second aspect the invention provides Lactococcus lactis subsp. lactis biovar diacetylactis CBS 130345 and Lactococcus lactis subsp. lactis biovar diacetylactis CBS 130347.

In a third aspect, the invention provides a fermented milk product obtainable by the process of the invention. Preferably the fermented milk product is Dahi or Curd. Dahi and Curd and other fermented milk products have been described in the background of the invention. The fermented milk product that may be produced by the process of the invention, preferable dahi or curd, comprises preferably at least 100 mg/l acetoin and/or at least 6 mg/l diacetyl. Preferably the acetoin is at least 150 mg/l, more preferably at least 200 mg/l, more preferably at least 250 mg/l, more preferably at least 300 mg/l more preferably at least 350 mg/l. Preferably, the diacetyl is produced at least 9 mg/l, more preferably at least 12 mg/l, more preferably at least 15 mg/l and more preferably at least 18 mg/l.

In a fourth aspect, the invention provides the use of a lactic acid bacterial strain that it is capable of producing at least 100 mg/l acetoin and/or at least 6 mg/l diacetyl in Test A for the production of diacetyl and/or acetoin in the process of the invention for the production of a fermented milk product, preferable dahi or curd, comprising preferably at least 1 00 mg/l acetoin and/or at least 6 mg/l diacetyl , preferably the acetoin produced is at least 150 mg/l, more preferably at least 200 mg/l, more preferably at least 250 mg/l, more preferably at least 300 mg/l more preferably at least 350 mg/l. Preferably, the diacetyl produced is at least 9 mg/l, more preferably at least 12 mg/l, more preferably at least 15 mg/l and more preferably at least 18 mg/l.

In a fifth aspect, the invention provides a method for the selection of a lactic acid bacterial strain that it is capable of producing at least 100 mg/l acetoin and/or at least 6 mg/l diacetyl in Test A subjecting a population of lactic acid bacteria to Test A.

YOGURT DEFINITIONS

Yogurt is a fermented milk product that contains the characteristic bacterial cultures *Lactobacillus bulgaricus* and *Streptococcus thermophilus*. All yogurt must contain at least 8.25% solids not fat. Full fat yogurt must contain not less than 3.25% milk fat, lowfat yogurt not more than 2% milk fat, and nonfat yogurt less than 0.5% milk. The full legal definitions for yogurt, lowfat yogurt and nonfat yogurt are specified in the Standards of Identity listed in the U.S. Code of Federal Regulations (CFR), in sections 21 CFR 131.200, 21 CFR 131.203, and 21 CFR 131.206, respectively.

The two styles of yogurt commonly found in the grocery store are set type yogurt and swiss style yogurt. Set type yogurt is when the yogurt is packaged with the fruit on the bottom of the cup and the yogurt on top. Swiss style yogurt is when the fruit is blended into the yogurt prior to packaging.

Ingredients

The main ingredient in yogurt is milk. The type of milk used depends on the type of yogurt – whole milk for full fat yogurt, lowfat milk for lowfat yogurt, and skim milk for nonfat yogurt. Other dairy ingredients are allowed in yogurt to adjust the composition, such as cream to adjust the fat content, and nonfat dry milk to adjust the solids content. The solids content of yogurt is often adjusted above the 8.25% minimum to provide a better body and texture to the finished yogurt. The CFR contains a list of the permissible dairy ingredients for yogurt.

Stabilizers may also be used in yogurt to improve the body and texture by increasing firmness, preventing separation of the whey (syneresis), and helping to keep the fruit uniformly mixed in the yogurt. Stabilizers used in yogurt are alginates (carageenan), gelatins, gums (locust bean, guar), pectins, and starch.

Sweeteners, flavors and fruit preparations are used in yogurt to provide variety to the consumer. A list of permissible sweeteners for yogurt is found in the CFR.

Bacterial Cultures

The main (starter) cultures in yogurt are *Lactobacillus bulgaricus* and*Streptococcus thermophilus*. The function of the starter cultures is to ferment lactose (milk sugar) to produce lactic acid. The increase in lactic acid decreases pH and causes the milk to clot, or form the soft gel that is characteristic of yogurt. The fermentation of lactose also produces the flavor compounds that are characteristic of yogurt. *Lactobacillus bulgaricus* and*Streptococcus thermophilus* are the only 2 cultures required by law (CFR) to be present in yogurt. Other bacterial cultures, such as *Lactobacillus acidophilus*, *Lactobacillus subsp. casei*, and Bifido-bacteria may be added to yogurt as probiotic cultures.Probiotic cultures benefit human health by improving lactose digestion, gastrointestinal function, and stimulating the immune system.

General Manufacturing Procedure

1. Adjust Milk Composition & Blend Ingredients: Milk composition may be adjusted to achieve the desired fat and solids content. Often dry milk is added to increase the amount of whey protein to provide a desirable texture. Ingredients such as stabilizers are added at this time.
2. Pasteurize Milk: The milk mixture is pasteurized at 185°F (85°C) for 30 minutes or at 203°F (95°C) for 10 minutes. A high heat treatment is used to denature the whey (serum) proteins. This allows the proteins to form a more stable gel, which prevents separation of the water during storage. The high heat treatment also further reduces the number of spoilage organisms in the milk to provide a better environment for the starter cultures to grow. Yogurt is pasteurized before the starter cultures are added to ensure that the cultures remain active in the yogurt after fermentation to act as probiotics; if the yogurt is pasteurized after fermentation the cultures will be inactivated.
3. Homogenize: The blend is homogenized (2000 to 2500 psi) to mix all ingredients thoroughly and improve yogurt consistency.

4. Cool Milk: The milk is cooled to 108°F (42°C) to bring the yogurt to the ideal growth temperature for the starter culture.
5. Inoculate with Starter Cultures: The starter cultures are mixed into the cooled milk.
6. Hold: The milk is held at 108°F (42°C) until a pH 4.5 is reached. This allows the fermentation to progress to form a soft gel and the characteristic flavor of yogurt. This process can take several hours.
7. Cool: The yogurt is cooled to 7°C to stop the fermentation process.
8. Add Fruit & Flavors: Fruit and flavors are added at different steps depending on the type of yogurt. For set style yogurt the fruit is added in the bottom of the cup and then the inoculated yogurt is poured on top and the yogurt is fermented in the cup. For swiss style yogurt the fruit is blended with the fermented, cooled yogurt prior to packaging.
9. Package: The yogurt is pumped from the fermentation vat and packaged as desired.

MATERIALS AND METHODS

Cultures

CBS number	Strain
CBS 130345	*Lactococcus lactis* subsp. *lactis* biovar. *diacetylactis* XM-165
CBS 130346	*Lactococcus lactis* subsp. *lactis* biovar. *diacetylactis* 100-D
CBS 130347	*Lactococcus lactis* subsp. *lactis* biovar. *diacetylactis* XM-152

All strains were deposited on 24 August 201 1 at the Centraalbureau voor Schimmelcultures (Fungal Biodiversity Centre), Uppsalalaan 8, 3508 AD, Utrecht, The Netherlands.

Test A

1 % (v/v) of a freshly prepared preculture that was grown overnight at 30°C in 10% (w/v) RSM (reconstituted skimmed milk powder) + 0.05% (w/v) yeast extract, was inoculated into fresh retail pasteurized milk and subsequently incubated for 24 hours at 40°C. As a control the test was carried out at 30°C. Diacetyl, acetoin and acetaldehyde concentrations in the fermentation samples were determined using the GC-FID method.

GC-FID method

Of each fermentation sample, 250 mg was accurately weighed into a suitable vial and 1 ml of internal standard solution (50 ppm 3-methyl-2-pentanone in acetonitrile) was accurately added. The sample was homogenized by vortexing and stored in a refrigerator for 30 min. The samples were subsequently centrifuged at 2000 rpm and the supernatant was further analysed. The quantification of the components acetaldehyde, diacetyl and acetoin in the fermentation samples was done by standard addition.

Standard solutions (solutions of a known concentration of analyte) were added to the sample (unknown solution). The initial concentration in the sample was determined by extrapolation . The analysis was done on an Agilent 7890 A GC equipped with a combi Pal liquid sampler, a split/splitless injector and an FID detector. The injector was operated in the in the split mode (split 1 :20) with an injection volume of 1 ìÉ and an injector temperature of 250 °C. The column used was a 30m x 0.25 mm i.d. fused-silica capillary column DB-624 (Jennings & Wilkins) with a stationary phase thickness of 1 .4 ìçç. Hydrogen was used as the carrier gas with a pressure of 100 kPa. The temperature of the column was initially set at 40°C (3 min hold), followed by a linear temperature gradient of 26.67 °C/min to 240 °C, at which value it was held for 3.5 min. The detector temperature was set at 250°C. Dionex Chromeleon software was employed for data processing.

3

Dairy Industry and Milk Products

Dairy products constitute one of the five main food groups that together comprise a healthy and balanced diet. Significant amounts of protein are found in dairy products along with most of the micronutrients we need. These include calcium, the B-group of vitamins, vitamin A, iodine, magnesium, phosphorus, potassium and zinc. Milk and dairy foods can also help to improve bone and dental health and may have a role to play in protecting the body against hypertension.

The progress in dairy science and technology has increased the level of hygiene required during the manufacture of dairy products in order to preserve their nutrients. The modern dairy industry requires the use of cleanable, corrosion-resistant stainless steel equipment to meet the needs of milk product consumers everywhere.

MILK PRODUCT ITEMS

There is but one real beverage and that is water. The other so-called beverages are foods, stimulants or sedatives. Milk is a rich food, one glass having as much food value as two eggs. Coffee, tea, chocolate and cocoa are stimulants, with sedative after-effects. Their food value depends largely on the amount of milk, cream and sugar put into them. Chocolate and cocoa are both drugs and foods. Alcohol is a stimulant at first, afterwards a sedative, and at all times an anaesthetic. When we think of drinking for the sake of supplying the bodily need of fluid, we should think of water and nothing else. If other liquids are taken, they should be taken as foods or drugs. Water is the best solvent known.

The alchemists of old spent much time and energy trying to find the universal solvent, believing that thereafter it would be easy to discover a method of making base metals noble. But they never found anything better than water. Water is the compound that in its various forms does most to change the earth upon which we live, and it is more necessary for the continuation of life than anything else except air. Pure water does not exist in nature, that is, we have never found a compound of the composition H_2O. Water always contains other matter.

The various salts are dissolved in it and it absorbs gases. The nearest we come to pure water is distilled. Pure water is an unsatisfied compound, and as soon as it is exposed it begins to absorb gases and take up salts and organic matter. Pure water differs from clean water. Clean or potable water is a compound which contains a moderate amount of salts, but very little of organic matter. Bacteria should be practically absent. Water that contains much of nitrogenous substances is unfit to use. If the water is very hard, heavily loaded with salts, it should not be used extensively as a drink, for if too much of earthy and mineral matter is taken into the system, the body is unable to get rid of all of them. The result is a tendency for deposits to form in the body. In places where the water is excessively charged with lime it has been noticed that the bones harden too early, which prevents full development of the body.

If the bones of the skull are involved, it means that there will not be room enough for the brain. Such diseases are rare in this country, but in parts of Europe they are not uncommon. If the water is very hard, a good plan is to distill it and then add a little of the hard water to the distilled water. People who partake of an excessive amount of various salts can perhaps drink distilled water to advantage, but those who take but a normal amount of the salts in their foods should have natural water. Water forms three-fourths of the human body, more or less. It is needed in every process that goes on within the body. "To be dry is to die."

Water keeps the various vital fluids in solution so that they can perform their function. Without water there would be no sense of taste, no digestion, no absorption of food, no excretion of debris, and hence no life. The water is the vehicle through which the nutritive elements are distributed to the billions of cells of the body, and it is also the vehicle which carries the waste to the various excretory organs.

We can live several weeks without food, but only a few days without water. Hot water and ice-cold water are both irritants. Water may be taken either warm or cool. It is best to avoid the extremes. The amount of water needed each twenty-four hours varies according to circumstances. Two quarts is a favourite prescription. Those who eat freely of succulent fruits and vegetables do not need as much as those who live more on dry foods. Salt in excess calls for an abnormal amount of water, for salt is a diuretic, robbing the tissues of their fluids and consequently more water has to be taken to keep up the equilibrium.

Naturally, more water is required when the weather is hot than when it is cool. On hot days warm water is more satisfying and quenches thirst more quickly than ice water. Warm water also stimulates kidney action, which is often sluggish in summer. Ice water is the least satisfactory of all, for the more one drinks the more he wants. A normal body calls for what water it needs, and no more. An abnormal body is no guide for either the amount of food or drink necessary. Many people do not like the taste of water, especially in the morning.

This means that the body is diseased. To a normal person cool water is always agreeable when it is needed, and it is needed in the morning. People with natural taste do not care for ice water, but other water is relished. The common habit of drinking with meals is a mistake. Man is the only animal that does this, and he has to pay dearly for such errors. Taking a bite of food and washing it down with fluid lead to undermastication and overeating, and then the body suffers from autointoxication. A mouthful of food followed by a swallow of liquid forces the contents of the mouth into the stomach before the saliva has the opportunity to act. The best way is to drink one or two glasses of water in the morning before breakfast. Partake of the breakfast, and all other meals, without taking any liquid. Sometimes there is a desire for a drink immediately after the meal is finished. If so, take some water slowly. If it is taken slowly a little will satisfy. If it is gulped down it may be necessary to take one or two glasses of water before being satisfied. Those who have a tendency to drink too much during warm weather will find very slow drinking helpful in correcting it. If there is any digestive weakness, the liquid taken immediately after a meal should be warm and should not exceed a cupful. Those with robust digestion may take cool water. Cold water chills the stomach. Digestion will not take place until the stomach has reached the temperature of about one hundred degrees Fahrenheit again, and if the stomach contents are chilled repeatedly the tendency is strong for the food to ferment pathologically, instead of being properly digested. For this reason it is not well to drink while there is anything left in the stomach to digest.

As stomach digestion generally takes two or three hours at least, it is well to wait this long before taking water after finishing a meal, and then drink all that is desired until within thirty minutes of taking the next meal. If the thirst should become very insistent before two or three hours have elapsed since eating, take warm water. Those who eat food simply prepared and moderately seasoned are not troubled much with excessive thirst. Two quarts of water daily should be sufficient for the adults under ordinary conditions. Here, as in eating, no exact amount will fit everybody. Make a habit of drinking at least a glass of water before breakfast, cleaning the teeth and rinsing the mouth before swallowing any, and then take what water the body asks for during the rest of the day. Taking too much water is not as injurious as overeating, but waterlogging the body has a weakening effect.

To drink with the meals is customary, not because it is necessary, but because we have a number of drinks which appeal to many people. Water is the drink par excellence. A food-beverage that is used by many is cambric tea, which is made of hot water, one-third or one-fourth of milk and a little sweetening. Children generally like this on account of the sweetness. It may be taken with any meal, when fluid is needed, but the amount should be limited to a cupful. It is not well to dilute the digestive juices too much.

The water taken in the morning helps to start the body to cleanse itself. Water drinking is a great aid in overcoming constipation. Constipated people generally overeat. Less food and more water will prove helpful in overcoming the condition. Unfortunately for the race, we have accustomed ourselves to partake of beverages containing injurious, poisonous substances. Inasmuch as this is the place to discuss the drugs contained in coffee and tea, we shall take the liberty of dwelling upon other habit-forming substances. They are all a part of the drug addictions of the race. For scientific discussion of these various substances we refer you to technical works. Coffee, tea and chocolate contain a poisonous alkaloid which is generally called caffeine. The theine in tea and the theobromine in cocoa are so similar to caffeine that chemists can not differentiate them. These drinks when first taken cause a gentle stimulation under which more work can be done than ordinarily, but this is followed by a reaction, and then the powers of body and mind wane so much that the average output of work is less than when the body is not stimulated.

The temporary apparently beneficial effect is more than offset by the reaction and therefore partaking of these beverages makes people inefficient. Coffee is very hard on the nerves, causing irritation, which is always followed by premature physical degeneration. Experiments of late indicate that children who use coffee do not come up to the physical and mental standard of those who abstain. The effect on the adults is not so marked because adults are more stable than children.

Those who are not used to coffee will be unable to sleep for several hours after partaking of a cup. Some people drink so much of it that they become accustomed to it. Coffee is not generally looked upon as one of the habit-forming drugs, but it is. However, of all the drugs which create a craving in the system for a repetition of the dose, coffee makes the lightest fetters.

It is surprising how often health-seekers inform the adviser that they "can not get along without coffee". If they would take a cup a few times a year, it would do no harm, but the daily use is harmful to all, even if they feel no bad effects and make it "very weak," which is a favourite statement of the women. Smoking, drinking beer and drinking coffee have a tendency to overcome constipation in those who are not accustomed to these things, but their action can not be depended upon for any length of time and the cure is worse than the disease. Tea drinking has much the same effect as coffee drinking, except that it is decidedly constipating. Perhaps this is because there is considerable of the astringent tannin in the tea leaves.

Chocolate is a valuable food. Those who eat of other aliments in moderation may partake of chocolate without harm, but if chocolate is used in addition to an excess of other food, the results are bad. The chocolate is so rich that it soon overburdens some of the organs of digestion, especially the liver. The Swiss consume much of this food and it is valuable in cases where it is necessary to carry concentrated rations.

Alcohol in some form seems to have been consumed by even very primitive people as far back as history goes. The Bible records an early case of intoxication from wine, and beer was brewed by the ancient Egyptians.

So much has been consumed that some people have a subconscious craving for it. There are cases on record where the very first drink caused an uncontrollable demand for the drug. Fortunately these cases are very rare. Alcohol is really not a stimulant, though it gives a feeling of glow, warmth and well-being at first, but this is followed by a great lowering of physical power, which gives rise to disagreeable sensations. Then the drinker needs more alcohol to stimulate him again. Then there is another depression with renewed demand: There is no end to the craving for the drug once it has mastered the individual. The lungs, heart, digestive organs, muscles, in fact, every structure in the body loses working capacity. Alcohol seems to have a special affinity for nervous tissue. A glass of beer or wine taken daily is no more harmful than a cup of coffee per day, but the coffee drinker does not make of himself such a public nuisance and menace as the man often does who drinks alcohol to excess. Formerly, it was respectable to drink. Some of our most noted public men were drunkards. Now a drunkard could not maintain himself in a prominent public position very long. To drink like a gentleman was no disgrace. Now real gentlemen do not get drunk.

In backward Russia they are becoming alarmed about the inroads of vodka, and are trying to decrease its consumption. France is trying to teach total abstinence to its young men because it disqualifies so many of them from military service to drink. Scandinavia is temperance territory. The German Kaiser has recently given a warning against drinking. The United States discourages drinking in the army and navy. Field armies are not supplied with alcoholics. Drinking is becoming disreputable.

It is very difficult to prove the harm done by excessive drinking of tea and coffee, also by the use of much tobacco, even if we do know that it is so. Everyone knows something about the deleterious effect of alcohol upon the consumer. Solomon wrote: "Wine is a mocker, strong drink is raging, and whosoever is deceived thereby is not wise. Who hath wounds without cause? Who hath redness of eyes?" Alcohol permanently impairs both body and mind. Depending on how much is taken, it may cause various ills, ranging from inflammation of the stomach to insanity. It reduces the power of the mind to concentrate and it diminishes the ability of the muscles to work. It reduces the resistance of the body and shortens life. Its first effect is to lull the higher faculties to sleep. Most drunkards do not recover from their disease, for drunkenness is a disease. The various drugs given to cure the afflictions are delusions. Strengthening the body, mind and the will and instilling higher ideals are the best methods of cure. Suggestive therapeutics, and the awakening of a strong resolve for a better life are powerful aids. Proper feeding should not be overlooked, for bad habits do not flourish in a healthy body.

Civilization necessitates self-control and considerable self-denial. Those who go in the line of least resistance are on the road to destruction. It is often necessary to overcome habits which produce temporary gratification of the senses. Warden Tynan of the Colorado Penitentiary, 96 per cent of the prisoners are brought there because they use alcohol. It is also well known that moral lapses are most common when the will is weakened through the use of liquor.

Those who have the welfare of the race at heart are therefore compelled to give considerable thought to this subject. Past experience, it will not help to try to legislate sobriety into the people. Education and industrialism are the factors which it seems to me will be most potent in solving the alcohol problem. Morality, which in the last analysis is a form of selfishness, will teach many that it is poor policy to reduce one's efficiency and thereby reduce the earning capacity and enjoyment of life.

More and more the employers of labour will realise that the use of alcohol decreases the reliability and worth of the worker. Many will take steps like the following: "In formal recognition of the fact, established beyond dispute by the tests of the new psychology, that industrial efficiency decreases with indulgence in alcohol and is increased by abstinence from it, the managers of a manufacturing establishment in Chester, Penn., have attacked the temperance problem from a new angle. "Unlike many railways and some other corporations, they do not forbid their employees to drink, but they offer 10 per cent advance in wages to all who will take and keep—the teetotaler's pledge. Incidentally, a breaking of the promise will mean a permanent severance of relations, but there is no emphasizing of that point, it being confidently expected that the advantage of perfect sobriety will be as well realised on one side as on the other."

Business has during the past two centuries been the great civilizer, the great moral teacher. It has found that honesty and righteousness pay and that injustice is folly. Business has led the way to the acceptance of a new ethics, and new morals. What has been said about alcohol applies to tobacco in a much smaller degree. The use of tobacco seems to lead to the use of alcohol. It retards the development of children. It is surely one of the causes of various diseases. Tobacco heart, sore throat and indigestion are well known to physicians. Tobacco contains one of the deadliest of poisons known. One-sixteenth of a grain of nicotine may prove fatal. The reason there are so few deaths from acute tobacco poisoning is that but very little of the nicotine is absorbed. Men who chew tobacco make themselves disagreeable to others. Smoking of cigarettes is to be condemned not only because it poisons the body, but causes inattention and inability to concentrate on the part of the smoker, as well.

Every little while he feels the desire to take a smoke, and if smoking is forbidden he devises means of getting away. He robs his employer of time for which he is paid and injures himself. The ability to work is decreased by indulgence in smoking. Recent experiments show that for a short time there is increased activity after a smoke, but the following depression is greater than the stimulation, so there is an actual loss.

A few years ago, according to Mr. Wilson, who was then Secretary of Agriculture, there were about 4,000,000 drug addicts or "dope fiends" in the United States. Without doubt this estimate was too high, for the proportion of addicts in the country is not as great as in the large cities. The drugs chiefly used are cocaine, opium, laudanum, morphine and heroin. These drugs are much more destructive than alcohol. Cocaine and heroin are the worst. It is very difficult to stop using any of them once the habit has been formed. Nearly every "fiend" dies directly or indirectly from the effect of his particular drug. Every one weakens the body so that there is not much resistance to offer to acute diseases. Every one destroys the will power so that a cure is exceedingly difficult.

It is well to bear in mind that all are not possessed of strong enough will power to resist their cravings and that some take to cocaine when they can not get liquor. Cocaine is far worse than alcohol. People should be very careful about taking patent medicines. There is no excuse for taking them. The most popular ones have as their basis one of the habit-forming drugs. Most of the soothing syrups contain opium in some form. To give babies opiates is a grave error, to speak mildly. It weakens the child, may lay the foundation for a deadly habit later in life, and often an overdose kills outright. Well informed mothers avoid such drugs and keep their children reasonably quiet by means of proper care. Many of the remedies for nasal catarrh and hay fever contain much cocaine. Cocaine is an astringent and a painkiller and people mistake the temporary lessening of discharge from the nose and disappearance of pain for curative effects. But there is nothing curative about it. In a short time the mucous membrane relaxes again and then the discharge is re-established. The nerves which were put out of commission resume their function and then the pain reappears.

Opium or one of its derivatives is generally present in the patent medicines given for coughs. Opium is also an astringent and will suppress secretions, but this is not a cure. Excessive secretions are an indication that the body is surcharged with poison and food. Let them escape and then live so that there will be internal cleanliness and then there will be no more coughs and colds. The unfortunate people who get into the habit of using these drugs degenerate physically, mentally and morally. They need more and more of their drug to produce the desired effect until they at last take enough daily to kill several normal men. Sometimes they are able to keep everybody in ignorance of what they are doing for years. They develop slyness and secretiveness. They become very suspicious. They are nearly always untruthful, and those who deal with them are surprised and wonder why those who used to be open and above-board now are furtive and dishonest. They often lie when there is not the slightest excuse for it.

The moral disintegration is often the first sign noticed. After habitually using any of these drugs for a while the body demands the continuation and if the victim is deprived of his accustomed portion there will be a collapse with intense suffering. Every tortured nerve in the body seems to call out for the drug. The victim will do anything to get his drug. He will lie, steal, and he may even attack those who are caring for him. For the time being he is insane. Many professional men use cocaine. It is a favourite with writers. It often shows in their work. Those who write under the inspiration of this drug often do some good work, but they are unable to keep to their subject. Their writings lack order. We have enough of such writings to have them classified as "cocaine literature". If there are 4,000,000, or even fewer, of these people in our land, it is a serious problem, for every one is a degenerate, to a certain degree. If the medical profession and the druggists would cooperate it would be easy enough to prevent the growth of a new crop of dope fiends. Of course, people would have to stop taking patent medicines, which often start the victims on the road to degeneration.

Then the physicians should stop prescribing habit-forming drugs, as well as all other drugs, and teach the people that physical, mental and moral salvation come through right living and right thinking. Unfortunately the medical profession is careless and is responsible for the existence of many of the drug addicts. A patient has a severe pain. What is the easiest way to satisfy him? To give a hypodermic injection of some opiate. The patient, not realizing the danger, demands a pain-killer every time he suffers. He soon learns what he is getting and then he goes to the drug store and outfits himself with a hypodermic outfit and drugs, and the first thing he knows he is a slave, in bondage for life.

This is no exaggeration. There are hundreds of thousands of victims to the drug habit who trace their downfall to the treatment received at the hands of reputable physicians, who do not look upon their practice with the horror it should inspire because it is so common. Doctors do not always bury their mistakes. Some of them walk about for years. In spite of laws against the sale of various drugs, they can be obtained. There are doctors and druggists of easy conscience who are very accommodating, for a price. There is no legitimate need for the use of one-hundredth of the amount of these drugs that is now consumed. A local injection of cocaine for a minor operation is justifiable, but none of the habit-forming drugs should be used in ordinary practice to kill pain, for the proper application of water in conjunction with right living will do it better and there are no evil after effects. Massage is often sufficient.

To show a little more clearly how some people become addicted to drugs, let us consider one of the latest, heroin: A few years ago this drug, which is an opium derivative, was practically unknown. It is much stronger than morphine and consequently the effect can be obtained more quickly by means of a smaller dose. Physicians thought at first that it was not a habit-forming drug, for they could use it over a longer period of time than they could employ morphine, without establishing the craving and the habit.

So they began to prescribe heroin instead of morphine, and many a morphine addict was advised to substitute heroin. All went well for a short while, until the victims found that they were enslaved by a drug that was even worse than morphine. Now, thanks chiefly to the medical profession, it is estimated that we have in our land several hundred thousand heroin addicts. Sallow of face, gaunt of figure, looking upon the world through pin-point pupils, with all of life's beauty, hope and joy gone, they are marching to premature death.

The medical profession furnishes more than its proportion of drug addicts. They know the danger of the drugs, but familiarity breeds contempt. If the public but knew how many of their medical advisers, who should always be clear-minded, are befuddled by drugs, there would be a great awakening. One eminent physician who has now been in practice about forty-five years and has had much experience with drug addicts, has said that according to his observations, about one physician in four contracts the drug habit. I believe this is exaggerated, but I am acquainted with a number of physicians who are addicts.

Physicians who smoke do not condemn the practice. Those who drink are likely to prescribe beer and wine for their patients. Those who are addicted to drugs use them too liberally in their practice. Those who have watched the effects of the various drugs, from coffee to heroin, must condemn their use. It is true that an occasional cup of coffee or tea, a glass of wine or beer does no harm. A cigarette a week would not hurt a boy, nor would on occasional cigar harm a man.

But how many people are willing to indulge occasionally? The rule is that they indulge not only daily, but several times a day,. and the results are bad. One bad habit leads to another, and the time always comes when it is a choice between disease and early death on one hand, and the giving up of the bad habits on the other, and when this time comes the bonds of habits are often so strong that the victim is unable to break them. We realise that knowledge will not always keep people out of temptation and that some individuals will take the broad way that leads to destruction in spite of anything that may be said. Youth is impatient of restraint and ever anxious for new experiences.

Regarding this serious matter of destructive drug use, much could be done by teaching people their place in society: That is, what they owe to themselves, their families and the public in general. In other words, teach the young people the higher selfishness, part of which consists of considerable self-control, self-denial and self-respect.

Drugs are too easy to obtain today. Some day people will be so enlightened that they will not allow themselves to be medicated. This is the trend of the times. Until such a time comes, society should protect itself by making it very difficult to get any of the habit-forming drugs. If necessary, the free hand of the physician should be stayed. Much of the confidence blindly given him is misplaced.

DAIRY INDUSTRY AND PROCESSING

The dairy industry has used membrane processing since its introduction in the food industry in the late 1960s to clarify, concentrate and fractionate a variety of dairy products. Applying membrane technology to whey processing allowed the production of refined proteins and commercial usage and thus transformed a waste byproduct from cheese production into a valuable product. In addition to whey processing, membrane technology is also used for fluid milk processing with clear advantages. Further, specific milk components can be obtained without causing a phase change to the fluid milk by the addition of heat as in evaporation, or an enzyme, as done in most cheese-making techniques.

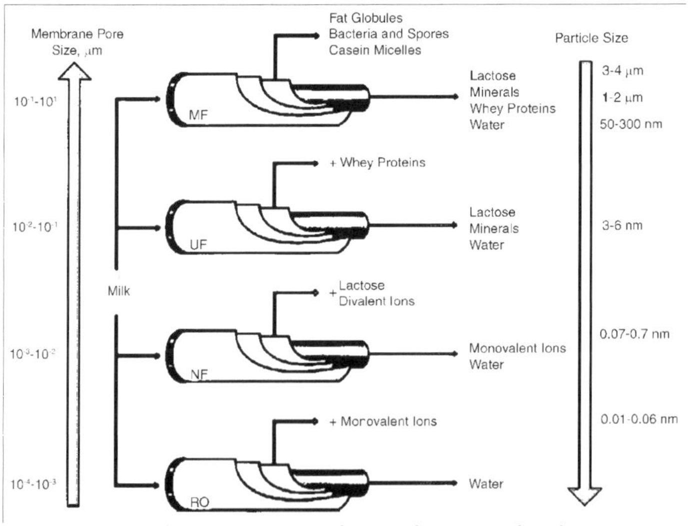

Fig. Milk Processing with Membrane Technology

The filtered milk can then be directly used in the manufacture of such dairy products as cheese, ice cream and yoghurt. By applying membranes with different pore sizes and molecular weight cut-offs (MWCOs), the milk can be modified by separating, clarifying, or fractionating a selected component in milk from other components. The pressure-driven membrane processes MF, UF, NFand RO are the most common membrane processes in the dairy industry and based on their applicability range it is possible to separate virtually every major component of milk as shown in Figure, thus enabling the manufacturing of products with unique properties and functionalities.

KEY MEMBRANE APPLICATIONS

The key applications of cross-flow membrane technology in the dairy industry are discussed.

Removal of Bacteria and Spores from Milk, Whey and Cheese Brine

The removal of bacteria and spores from milk to extend its shelf-life by MF is an alternative way to ultrapasteurization. In this approach, the organoleptic and chemical properties of the milk are unaltered. The first commercial system of this so-called Bactocatch was developed by Alfa Laval [1–3] and marketed by Tetra Pak under the name Tetra Alcross®Bactocatch. In this process, the raw milk is separated into skim milk and cream. The resulting skim milk is microfiltered using ceramic membranes with a pore size of 1.4 μm at constant transmembrane pressure (TMP).

Thus, the retentate contains nearly all the bacteria and spores, while the bacterial concentration in the permeate is less than 0.5 per cent of the original value in milk.

The retentate is then mixed with a standardized quantity of cream. Subsequently, this mix is subjected to a conventional high heat treatment at 130°C for 4s and reintroduced into the permeate, and the mixture is then pasteurized. Since less than 10% of the milk is heat treated at the high temperature, the sensory quality of the milk is significantly improved.

MF for the removal of bacteria and spores can be further applied in the production of other dairy products. In the production of cheese, the use of low bacterial milk improves also the keeping quality of cheese due to the removal of spores, thus eliminating the need of additives (*e.g.*, nitrate). While in the production of whey protein concentrates (WPC) and isolates (WPI), this MF concept is used to remove bacteria and spores giving a high quality product. Hence, by applying MF the heat treatment of the WPC/WPI is kept to a minimum, which preserves the functional properties of the whey proteins

Finally, in the manufacture of cheese the concentrated curd is submerged in a salt solution to improve the cheese preservation and to develop the flavour and other cheese properties. This process is called brining. Efficient sanitation of cheese brine has become a major concern to the dairy industry in recent years. This results from the possibility of post-contamination of cheeses in the brine, especially by pathogenic bacteria..

The application of MF for sanitation of cheese brine, using ceramic or spiralwound membranes, results in a superior cheese quality compared to the traditional processes of heat treatment and kieselguhr filtration. MFhas the advantages of being simple to perform, of maintaining the chemical balance of the brine and of eliminating filter aids. In the brine treatment by MF it is normally necessary to make a prefiltration of the brine solution, which is easily done by dead-end filter bag or cartridge with a pore size of 100 μm.

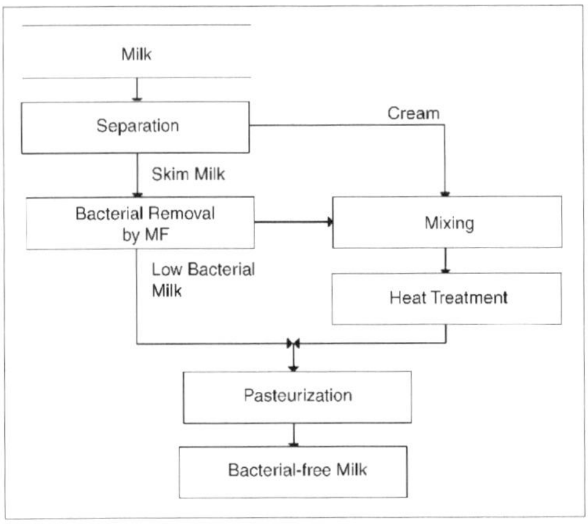

Fig. Bacterial Removal from Milk by MF

Milk Protein Standardization, Concentration and Fractionation

The protein content of milk is subjected to natural variations during the year. Standardization of milk by UF offers the possibility of increasing or decreasing the protein content in milk without the need of adding milk powders, casein and whey protein concentrates. Skim milk and 1% milk with increased protein content have an improved appearance (whiter milk) and higher viscosity.

The sensory quality of increased protein milk is therefore more similar to that of higher fat milks resulting in an improved consumer appeal. Another application of UFis the standardization of protein and total solids in milk for use in fermented dairy products, such as cream cheeses, yoghurt and cottage cheeses.

The resulting dairy products have superior quality and sensory characteristics compared to those produced from milk concentrated by conventional methods. With the quality obtained by membrane filtration, attributes such as consistency, post-processing and extent of syneresis are easier to control. However, the use of membrane-processed milk often requires an adjustment in starter culture selection and fermentation conditions due to the compositional changes in the UF milk. Concentrationof milk, whichconventionally isdonebyevaporation techniques, can also be achieved by RO. The concentrated milk has its greatest potential in ice-cream manufacturing, since all the solids are retained in the concentrateand 70% of the water is removed. MF and/or UF are used in the production of milk protein concentrates (MPC), which are products containing 50–58% of protein.

These products are used as food additives and it is therefore extremely important to maintain the functionality of the proteins. By using UF membranes in combination with MF and/or diafiltration (DF) with the corrected adjustments of pH, temperature and filtration conditions, it is possible to produce the desirable MPC for a specific food application. Themost promisingMFapplication in the dairy industry is the fractionation of milk protein.

The separation of micellar casein from the whey proteins can be achieved by ceramic membranes with a pore size of 0.2 μm at a constant TMP. The resulting retentate has a high concentration of native calcium phosphocaseinate that can be used for cheese making. Native casein has an excellent rennet-coagulation ability that will make calcium phosphocaseinate an exceptional enrichment for cheese-milk. The permeate can be further processed byUFto produce high-qualityWPC. These protein concentrates can be further separated into lactoferrin, β-lactoglobulin and α-lactalbumin via ion-exchange chromatography. Both β-lactoglobulin and α-lactalbumin have great potential markets.

β-lactoglobulin can be used as a gelling agent and α-lactalbumin, which is very rich in tryptophan, can be used in the production of peptides with physiological properties.

Another application can be the production of infant milk. The fractionation of milk proteins using membrane technology enables the recovery of value-added protein ingredients. Further, the casein and whey proteins are separated without the need of heat or enzymes. The potential applications of membrane separation in milk processing are shown in Figure.

Whey Protein Concentration and Fractionation

Whey is a by-product from the cheese industry. It has low content of solids and high biological oxygen demand (BOD), which creates a major disposal problem for the dairy industry. In the past, all whey was disposed of as sewage, sprayed on fields or used for animal feed.

By applying membrane technology whey can be concentrated to produce WPC and WPI, as well as fractionated and purified to obtain purified α-lactalbumin and β-lactoglobulin. Hence, a once wasted product can be converted into high value-added products and at the same time one of the key pollution problems of the dairy industry can be solved. Consequently, the use of UFand RO to concentrate whey was one of the first applications of membranes in the dairy industry. Due to the complexity and diversity of whey, it is necessary to use different membrane processes to produce a specific product.

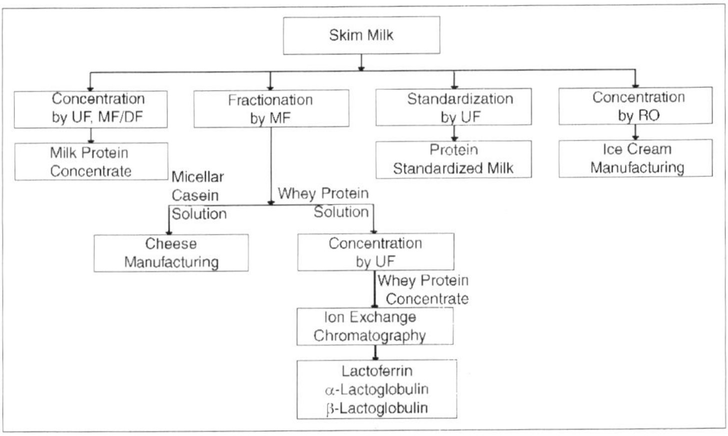

Fig. Applications of Membrane Technology in Milk Processing

The production of WPC with 35–85% protein in the total solids can be achieved by a combination of UF and DF. MF can be used as a pretreatment to remove both bacteria and fat and allows the production of WPI with 90% protein in the total solids. Whey proteins have not only a high nutritional value but also functional properties. They can be used as gelling, emulsifying and foaming agents. Therefore, whey concentrates have farreaching applications not only in dairy foods, but also in confectionary, nutritional foods, beverages and even processed meats.

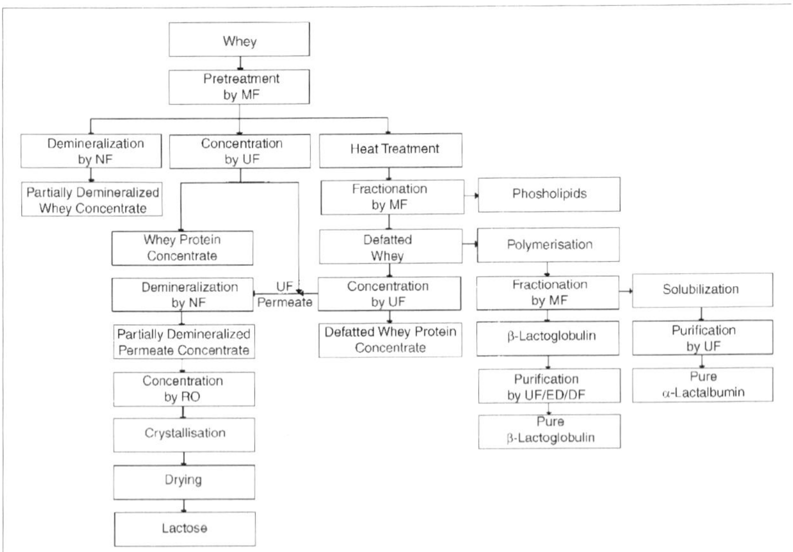

Fig. Applications of Membrane Technology in Whey Processing

The presence of fat in whey leads to decreased functional properties and shorter storage time. Several processes involving membranes have been developed to remove the residual fat from whey. The most common process, developed byMaubois *et al.* and Fauquant *et al.*, exploits the ability of the phospholipids to aggregate by calcium binding under moderate heat treatment for 8 min at 50°C. This process is called thermocalcic precipitation. Defatted whey is then obtained by MF with a pore size of 0.14 μm to separate the resulting precipitate. Defatted whey can be further processed by UF, which also improves the performance in the subsequent membrane processes. The defatted WPC has a foaming capacity similar to that of egg white and the same protein content. Its applications can be as raw material in the pastry and icecream production. The MF retentate, which contains a high amount of phospholipids, can be used as an effective emulsifier agent for food and cosmetic applications.

The purified proteins β-lactoglobulin and α-lactalbumin can be obtained from the defatted whey. At low pH (4.0–4.5) and under moderate heat treatment for 30 min at 55°C, α-lactalbumin polymerizes reversibly entrapping most of the residual lipidsand the other whey proteins with the exception of the β-lactoglobulin. The fractionation of β-lactoglobulin from the remaining proteins can then be done by MF with apore size of 0.2 μm or centrifugation. The resulting soluble phase, rich in β-lactoglobulin, can be further purified by UF coupled with electrodialysis (ED) or DF. Purification of α-lactalbumin from the MF retentate can be achieved by solubilization at a neutral pH and subsequently by UF using a membrane with an MWCO of 50,000 Dalton.

It has also been reported that membranes can be applied for the isolation of Kcasein-glycomacropeptid (GMP) from cheese whey. GMP can find several applications in the pharmaceutical industry. Studies have shown that GMP avoids the adhesion of Escherichia coli cells to the intestine walls, protects against influenza and prevents adhesion of tartar to teeth. It should also be noted that membrane filtration also plays a major role in the lactose manufacture from whey using UF and RO and in the production of lowcarbohydrate beverages with high dairy protein content.

Whey Demineralization

In the dairy industry, the NFprocess is used to concentrate and partially demineralize liquid whey. Due to the selectivity of the membranes most of the monovalent ions, the organic acids, and some of the lactose will pass the membrane. NF is a very interesting alternative to ion exchange and ED if moderate demineralization is required. One advantage of NF compared to the other two processes is that NF is a simple process, which partially demineralizes and concentrates the whey at the same time.

The maximum level of demineralization by NF is about 35% reduction of the ash content with a concentration factor of about 3.5–4. By applying a DF step it is possible to increase the level of demineralization up to 45%. Other applications of NF in whey processing include: concentration and partial demineralization of whey UF permeates prior to the manufacture of lactose and lactose derivatives, converting. salt whey. to normal whey while solving a disposal problem, treating cheese brine solutions to be reused.

Cheese Manufacturing

Another early application of membrane technology in the dairy industry was in cheese manufacturing for production of Feta cheese and brine treatment by UF. Nowadays, membrane-processed milk is also successfully used in the manufacturing of quark and cream cheeses. Together with WPC production, the use of UF milk for the production of cheese is the most widespread application of membranes in the dairy industry.

The advantages of UFconcentrated milk in cheese making compared to traditional methods are the following:

- Increases the total solids, which increases the cheese yield and therefore decreases the production costs in terms of energy and equipment;
- Reduces the rennet and starter culture requirements since UF-milk has a good ability of enzymatic coagulation;
- Reduces the wastewater processing costs of the cheese plant;
- Improves the quality and composition control;
- Increases the nutritional value due to the incorporation of the whey protein in the cheese.

UF in cheese processing can be used in three ways:

1. *Preconcentration*: The standardized cheese milk is concentrated by a factor of 1.2–2 and it can be used for most cheese types. This allows the capacity of the cheese vats and whey draining equipment to be doubled. However, the cheese yield will not be significantly improved since only 4. 5–5% of the protein content is increased. It is used to produce Cheddar, Cottage Cheese and Mozzarella, and it can be used to standardize cheese milk and manipulate its mineral composition, resulting in a more consistent quality in the final product.
2. *Partial Concentration*: The standardized cheese milk is concentrated by a factor 2–6. It is used in the manufacture of Cheddar cheese by using for example, the APV-SiroCurd process, in which the milk is concentrated five times with DF in order to standardize the salt balance. It is also used to produce other cheese types like Queso Fresco, structure Feta, Camembert and Brie.
3. *Total Concentration*: The standardized cheese milk is concentrated to the total solids content in the final cheese. This provides the maximum yield increase and since there is no whey drainage, the cheese can be manufactured without the need for a cheese vat. It is

used to produce cast Feta, quark, cream cheese, Ricotta and Mascarpone.

The UF permeate, which contains mainly lactose, can be concentrated by RO. The permeate from the RO process can be polished by another RO unit. After pasteurization orUVlight treatment, the permeate from the polisher can be used at the plant as process water, thus reducing the water costs of the plant. Although UF has advantages in cheese production, the increase of whey content in the cheesedueto the concentration of all milk proteinscan have a negative effectonthe ripening of semihard and hard cheeses. Therefore, UFshould be viewed as a complementary process to cheese manufacturing and not as an alternative process.

COWS USED FOR THEIR MILK

Cows produce milk for the same reason that humans do: to nourish their young. In order to force the animals to continue giving milk, factory farm operators typically impregnate them using artificial insemination every year. Calves are generally taken from their mothers within a day of being born—males are destined for veal crates or barren lots where they will be fattened for beef, and females are sentenced to the same fate as their mothers.

After their calves are taken away from them, mother cows are hooked up, several times a day, to milking machines. These cows are genetically manipulated, artificially inseminated, and often drugged to force them to produce about four and a half times as much milk as they naturally would to feed their calves.

Animals are often dosed with bovine growth hormone (BGH), which contributes to a painful inflammation of the udder known as "mastitis." (BGH is used widely in the U.S. but has been banned in Europe and Canada because of concerns over human health and animal welfare.) According to the industry's own figures, between 30 and 50 per cent of dairy cows suffer from mastitis, an extremely painful condition.

A cow's natural lifespan is about 25 years, but cows used by the dairy industry are killed after only four or five years. An industry study reports that by the time they are killed, nearly 40 per cent of dairy cows are lame because of the intensive confinement, the filth, and the strain of being almost constantly pregnant and giving milk. Dairy cows' bodies are turned into soup, companion animal food, or low-grade hamburger meat because their bodies are too "spent" to be used for anything else.

VEAL CALVES

Male calves—"byproducts" of the dairy industry—are generally taken from their mothers when they are less than 1 day old. Many are shipped off to barren, filthy feedlots to await slaughter. Others are kept in dark, tiny crates where they are kept almost completely immobilized so that their flesh stays tender. In order to make their flesh white, the calves are fed a liquid diet that is low in iron and has little nutritive value.

This heinous treatment makes the calves ill, and they frequently suffer from anemia, diarrhea, and pneumonia. Frightened, sick, and alone, these calves are killed after only a few months of life so that their flesh can be sold as veal. All adult and baby cows, whether raised for their flesh or their milk, are eventually shipped to a slaughterhouse and killed.

The good news is that removing dairy products from your diet is easier than ever. Today there is a multitude of nondairy "dairy" products on the market, such as soy, rice, and almond milk and soy ice cream. Check out a list of our favourite dairy and meat alternatives.

IMPORTANT SOURCE OF SUBSIDIARY INCOME TO SMALL/ MARGINAL FARMERS

Dairying is an important source of subsidiary income to small/marginal farmers and agricultural labourers. The manure from animals provides a good source of organic matter for improving soil fertility and crop yields. The gober gas from the dung is used as fuel for domestic purposes as also for running engines for drawing water from well. The surplus fodder and agricultural by-products are gainfully utilised for feeding the animals.

Almost all draught power for farm operations and transportation is supplied by bullocks. Since agriculture is mostly seasonal, there is a possibility of finding employment throughout the year for many persons through dairy farming. Thus, dairy also provides employment throughout the year. The main beneficiaries of dairy programmes are small/marginal farmers and landless labourers. A farmer can earn a gross surplus of about Rs. 12,000 per year from a unit consisting of 2 milking buffaloes.

The capital investment required for purchase of 2 buffaloes is Rs. 18,223/-. Even after paying a sum of Rs. 4294/- per annum towards repayment of the loan and interest the farmer can earn a net surplus of Rs. 6000 - 9000/- approximately per year. Even more profits can be earned depending upon the breed of animal, managerial skills and marketing potential. World Bank estimates about 75 per cent of India's 940 million people are in 5.87 million villages, cultivating over 145 million hectares of cropland. Average farm size is about 1.66 hectares. Among 70 million rural households, 42 per cent operate upto 2 hectares and 37 per cent are landless households.

These landless and small farmers have in their possession 53 per cent of the animals and produce 51 per cent of the milk. Thus, small/marginal farmers and land less agricultural labourers play a very important role in milk production of the country. Dairy farming can also be taken up as a main occupation around big urban centres where the demand for milk is high.

SCOPE FOR DAIRY FARMING AND ITS NATIONAL IMPORTANCE

The total milk production in the country for the year 2001-02 was estimated at 84.6 million metric tonnes. At this production, the per capita availability was to be 226 grams per day against the minimum requirement of 250 grams per day as recommended by ICMR. Thus, there is a tremendous scope/potential for increasing the milk production. The population of breeding cows and buffaloes in milk over 3 years of age was 62.6 million and 42.4 million, respectively (1992 census). Central and State Governments are giving considerable financial assistance for creating infrastructure facilities for milk production. The nineth plan outlay on Animal Husbandry and Dairying was Rs. 2345 crores.

FINANCIAL ASSISTANCE AVAILABLE FROM BANKS/NABARD FOR DAIRY FARMING

NABARD is an apex institution for all matters relating to policy, planning and operation in the field of agricultural credit. It serves as an apex refinancing agency for the institutions providing investment and production credit. It promotes development through formulation and appraisal of projects through a well organised Technical Services Department at the Head Office and Technical Cells at each of the Regional Offices.

Loan from banks with refinance facility from NABARD is available for starting dairy farming. For obtaining bank loan, the farmers should apply to the nearest branch of a commercial or co-operative Bank in their area in the prescribed application form which is available in the branches of financing banks. The Technical Officer attached to or the Manager of the bank can help/give guidance to the farmers in preparing the project report to obtain bank loan.

For dairy schemes with very large outlays, detailed reports will have to be prepared. The items of finance would include capital asset items such as purchase of milch animals, construction of sheds, purchase of equipments etc. The feeding cost during the initial period of one/two months is capitalised and given as term loan. Facilities such as cost of land development, fencing, digging of well, commissioning of diesel engine/pumpset, electricity connections, essential servants' quarters, godown, transport vehicle, milk processing facilities etc. can be considered for loan. Cost of land is not considered for loan. However, if land is purchased for setting up a dairy farm, its cost can be treated as party's margin upto 10% of the total cost of project.

SCHEME FORMULATION FOR BANK LOAN

A Scheme can be prepared by a beneficiary after consulting local technical persons of State animal husbandry department, DRDA, SLPP etc., dairy co-operative society/union/federation/commercial dairy farmers. If possible, the beneficiaries should also visit progressive dairy farmers and government/ military/agricultural university dairy farm in the vicinity and discuss the profitability of dairy farming.

A good practical training and experience in dairy farming will be highly desirable. The dairy co-operative societies established in the villages as a result of efforts by the Dairy Development Department of State Government and National Dairy Development Board would provide all supporting facilities particularly marketing of fluid milk.

Nearness of dairy farm to such a society, veterinary aid centre, artificial insemination centre should be ensured. There is a good demand for milk, if the dairy farm is located near urban centre.

The scheme should include information on land, livestock markets, availability of water, feeds, fodders, veterinary aid, breeding facilities, marketing aspects, training facilities, experience of the farmer and the type of assistance available from State Government, dairy society/union/federation.

The scheme should also include information on the number of and types of animals to be purchased, their breeds, production performance, cost and other relevant input and output costs with their description.

Based on this, the total cost of the project, margin money to be provided by the beneficiary, requirement of bank loan, estimated annual expenditure, income, profit and loss statement, repayment period, etc. can be worked out and shown in the Project report. A format developed for formulation of dairy development schemes.

SCRUTINY OF SCHEMES BY BANKS

The scheme so formulated should be submitted to the nearest branch of bank. The bank's officers can assist in preparation of the scheme for filling in the prescribed application form. The bank will then examine the scheme for its technical feasibility and economic viability.

Technical Feasibility—this would briefly include:

- Nearness of the selected area to veterinary, breeding and milk collection centre and the financing bank's branch.
- Availability of good quality animals in nearby livestock market. The distribution of important breeds of cattle and buffaloes. The reproductive and productive performance of cattle and buffalo breeds.
- Availability of training facilities.
- Availability of good grazing ground/lands.
- Green/dry fodder, concentrate feed, medicines etc.
- Availability of veterinary aid/breeding centres and milk marketing facilities near the scheme area.

Economic Viability—this would briefly include:

- *Unit Cost*: The average unit cost of dairy animals for some of the States.
- Input cost for feeds and fodders, veterinary aid, breeding of animals, insurance, labour and other overheads.
- Output costs i.e. sale price of milk, manure, gunny bags, male/female calves, other miscellaneous items etc.
- Income-expenditure statement and annual gross surplus.
- Cash flow analysis.
- Repayment schedule (i.e. repayment of principal loan amount and interest).

Other documents such as loan application forms, security aspects, margin money requirements etc. are also examined. A field visit to the scheme area is undertaken for conducting a techno-economic feasibility study for appraisal of the scheme. Model economics for a two animal unit and mini dairy unit with ten buffaloes.

STAINLESS STEEL A SUSTAINABLE MATERIAL

Before we can determine whether stainless steel is a sustainable material, we should first define what we mean by sustainability in relation to what is known as the triple bottomline: People, Planet and Profit.

PEOPLE

The material, in its use or in its production process, respects the human being, especially in terms of health and safety. A sustainable material does not harm the people working to produce it, or the people who handle it during its use, recycling and ultimate disposal. Stainless steel is not harmful to people during either its production or use. A protective layer forms naturally on all stainless steels because of the inclusion of chromium. The passive layer protects the steel from corrosion–ensuring a long life. As long as the correct grade of stainless is selected for an application, the steel remains inert and harmless to the people who handle it and the environment. These characteristics have made stainless steel the primary material in medical, food processing, household and catering applications.

PLANET

The emission footprints of the material, especially those related to carbon, water and air, are minimised. Reuse and recyclability are at high levels. The material has low maintenance costs and a long life, both key indicators that the impact of the material on the planet is at the lowest levels possible. The electric arc furnace (EAF), the main process used to make stainless steels, is extremely efficient. An EAF has a low impact on the environment in terms of both CO2 and other emissions.

The EAF is also extremely efficient at processing scrap stainless, ensuring that new stainless steel has an average recycled content of more than 60%. Stainless steels are easily recycled to produce more stainless steels and this process can be carried on indefinitely. It is estimated that about 80% of stainless steels are recycled at the end of their life. As stainless steel has a high intrinsic value, it is collected and recycled without any economic incentives from the public purse.

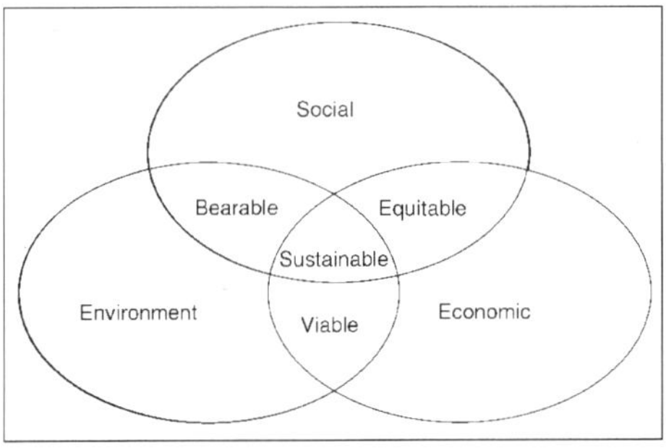

PROFIT

The industries producing the material show long-term sustainability and growth, provide excellent reliability and quality for their customers, and ensure a solid and reliable supply-chain to the end consumer.

Choosing stainless steel for an application ensures that it will have low maintenance costs, a long life and be easy to recycle at the end of that life. This makes stainless an economical choice in consumer durables (such as refrigerators and washing machines) and in capital goods applications (such as transportation, chemical and process applications).

Stainless steels also have better mechanical properties than most metals. Its fire and corrosion resistance make stainless a good choice in transportation, building or public works such as railways, subways, tunnels and bridges. These properties, together with stainless steels' mechanical behaviour, are of prime importance in these applications to ensure human beings are protected and maintenance costs are kept low. Stainless also has an aesthetically pleasing appearance, making it the material of choice in demanding architectural and design projects.

Taking into account its recyclability, reuse, long life, low maintenance and product safety, the emissions from the production and use of stainless steels are minimal when compared to any other alternative material.

A detailed and precise analysis of the sustainability of stainless steel makes the choice of stainless a logical one. This might explain why, as society and governments are becoming more conscious of environmental and economic factors, the growth in the use of stainless steel has been the highest of any material in the world.

DAIRY PRODUCTS

The dairy products vary greatly. Some cows give richer milk than others. Butter may be almost pure fat, or it may contain much water and salt. The cheeses are rich or poor in protein and fats according to method of making. Cottage cheese may be well drained or quite watery. Therefore, this table gives only approximate contents.

	Water	Pro-tein	Fat	Carbohy-drates	Ash	Calories Per lb.
Whole milk	87.00	3.3	4.0	5.0	0.7	325
Cream	74.00	2.5	18.5	14.5	0.5	910
Buttermilk	91.00	3.0	0.5	4.8	0.7	165
Butter	...	..	82.4	..	..	3475
Cheese, whole milk	33.70	26.0	34.2	2.3	3.8	1965
"skimmed milk	45.70	31.5	16.4	2.2	4.2	1320

Milk is not a beverage. It is a food. A quart of milk contains as much food and fuel value as eight eggs or twelve ounces of lean beef. That is, a cupful is equal to two eggs or three ounces of lean beef. This shows that milk should not be taken to quench thirst, but to supply nourishment. Milk is one of our most satisfactory and economical albuminous foods, even at the present high prices. In many foods from 5 to 10 per cent of the protein goes to waste. In milk the waste does not ordinarily amount to more than about 1 per cent.

This fluid generally leaves the stomach within one or one and one-half hours after being ingested. In spite of its merits as a food some writers on dietetics advocate that adults stop using it, giving it only to the young. Milk is an excellent food when properly used. When abused it tends to cause discomfort, disease and death, and so does every other food known to man. Milk is given in fevers and in other diseases, when the digestive and assimilative processes are suspended. This is a serious mistake and has caused untold numbers of deaths. When the digestion has gone on a strike all feeding is destructive.

Milk and meat broths, which are generally given, are about the worst foods that could be selected under the circumstances, for they decay very easily, and are excellent food for the numerous bacteria that thrive in the digestive tract during disease. These foods must decay when they are not digested, for the internal temperature of the body during fevers is over one hundred degrees Fahrenheit. When bacteria are present in excess they give off considerable poison, which makes the patient worse. If circumstances are such that it is necessary to feed during acute disease, which is always injurious to the patient, let the food be the least harmful obtainable, such as fruit juices.

Even they do harm. In our country cow's milk is used almost exclusively. In other lands the milk of the mare, the ass, the sheep, the goat and of other animals is used. The objection voiced against cow's milk is that it is an unnatural food for man, only fit for the calf, which is equipped with several stomachs and is therefore able to digest the curds which are larger and tougher than the curds formed from human milk. It is said that the curds of cow's milk are so indigestible that the human stomach cannot prepare them for entry into the blood. This is probably true, but it is also true of other protein-bearing foods. The digestion and assimilation of proteins are begun in the stomach and completed in the intestines, and the protein in milk is one of the most completely utilized of all proteins.

To call a food unnatural means nothing, for we can call nearly all foods unnatural and defend our position. A natural food is presumably a nutritious and digestible aliment that is produced in the locality where it is consumed, one that can be utilized without preparation or preservation. So we may say that a resident of New York should not use figs, dates, bananas and other products of tropical and semi-tropical climates, for they are not natural in the latitude of New York. We can take the position that it is unnatural for people to eat grains, which need much grinding, for the birds are the only living beings supplied with mills.

We can further say that it is unnatural to eat all cooked and baked foods. But such talk is not helpful. The more a person uses his brain the less power he has left for digestion and therefore it is necessary to prepare some of the foods so that they will be easy to digest. Man is such an adaptable creature that we are not sure what he subsisted on before he became civilized and are therefore unable to say what his natural food is. We know that in the tropics fruits play an important part in nourishing savages, while in the frozen north fat flesh is the chief food. Perhaps there is no natural food for man. Some of those who advocate the disuse of milk have a substitute or imitation to take its place, nut milk made from finely ground nuts and water. Like all other imitations, it is inferior to the original. It is more difficult to digest than real milk and the flavour is quite different. The objection that milk is indigestible is not borne out by the experience of those who give it under proper conditions. It is true that milk disagrees with a few, but so do such excellent foods as eggs, strawberries and Concord grapes, and many other aliments which are not difficult to digest. This is a matter of individual peculiarity. Some can take boiled milk, but are unable to take it fresh, and *vice versa*. Outside of the few exceptions, milk digests in a reasonable time and quite completely. It is easier to digest than the legumes which are rich in protein.

It is also easier to digest than nuts, which contain much protein. The milk sugar causes no trouble and cream is one of the easiest forms of fat to digest, if taken in moderation. The protein in milk will cause no inconvenience if the milk is eaten slowly, in proper combinations and not to excess. The rennet in the stomach curdles the casein. The hydrochloric acid and the pepsin in the gastric juice then begin to break down and dissolve the clots, and the process of digestion is completed in the small intestines. Those who overeat of milk in combination with other foods will derive benefit from omitting the milk. They will also be benefitted if they continue using milk and omit either the starch or the meat. When foods disagree, in nearly every instance it is due to the fact that too much has been eaten and too many varieties partaken of at a meal. Some may single out the milk or the meat as the offenders.

Others may point to the starches, and still others to the vegetables with their large amount of indigestible residue. They are all right and all wrong, for all the foods help to cause the trouble. However, such reasoning does not solve the problem. If the meals cause discomfort and disease, reduce the amount eaten, take fewer varieties at a meal and simplify the cooking. Those who eat simple meals and are moderate are not troubled with indigestion. Those who eat such mushy foods as oatmeal and cream of wheat usually take milk or cream and sugar with them. This should not be done, for such dressing stimulates the appetite and leads to undermastication. Neither children nor adults chew these soft starchy foods enough. The result is that the breakfast ferments in the alimentary tract. After a few months or years of such breakfasts, some kind of disease is sure to develop.

Mushy starches dressed with rich milk and sugar are responsible for a large per cent of the so-called diseases of children, which are primarily digestive disturbances. Colds, catarrhs and adenoids are, of course, due to improper eating extending over a long period of time.

Nothing should be eaten with mushy starches except a little butter and salt. After enough starch has been taken, a glass of milk may be eaten. If parents would only realise that they are jeopardizing the health and lives of their dear ones when they feed them habitually on these soft messes, which ferment easily, there would be a remarkable decrease in the diseases of childhood and in the disgraceful infant and childhood mortality, for several hundred thousand children perish annually in this country.

Milk is often referred to as a perfect food, and it is the perfect food for infants. The young thrive best on the healthy milk given by a female of their own species. Every baby should be fed at the breast. The milk contains the elements needed by the body. The ash is composed of the various salts necessary for health, containing potassium, chlorine, calcium, magnesium, iron, silicon and other elements. For the nourishment of the body we need water, protein, fat, carbohydrates and salts, so it will be seen that milk is really a complete food. However, as the body grows the nutritive requirements change and milk is therefore not a balanced food for adults.

It may be interesting to note that there is no starch in milk and that infants fed at the breast exclusively obtain no starchy food. Many babies get no starch for nine, ten or even twelve months, and this is well, for they do not need it. They grow and flourish best without it. Milk is an emulsion. It is made up of numerous tiny globules floating in serum. The size of the globules varies, but the average is said to be about 1/10,000 of an inch in diameter. These globules are fatty bodies. There are other small bodies, containing protein and fat, which have independent molecular movement. The milk is a living fluid. When it is tampered with it immediately deteriorates. Without doubt, nature intended that the milk should go directly from the mammary gland into the mouth of the consumer, but this is not practicable when we take it away from the calf. However, if we are to use sweet milk it is best to consume it as nearly like it is in its natural state as possible.

It is quite common to drink milk rapidly. This should not be done. Take a sip or a spoonful at a time and move it about in the mouth until it is mixed with saliva. It is not necessary to give it as much mouth preparation as is given to starchy food. If it is drunk rapidly like water large curds from in the stomach. If it is insalivated it coagulates in smaller curds and is more easily digested, for the digestive juices can tear down small soft curds more easily than the large tough ones. Milk should not form a part of any meal when other food rich in protein is eaten. Our protein needs are small, and it is easy to get too much. Whole wheat bread and milk contain all the nourishment needed. On such a diet we can thrive indefinitely. This is information, not a recommendation. The bread should be eaten either before or after partaking of the milk. Do not break the bread into the milk. If this is done, mastication will be slighted. Bread needs much mastication and insalivation. When liquid is taken with the bread, the saliva does not flow so freely as when it is eaten dry.

Fruit and milk make a good combination, but no starchy foods are to be taken in this meal. Take a glass of milk, either sweet or sour, and what fruit is desired, insalivating both the fruit and the milk thoroughly. If you have read that the combination of fruit and milk has proved fatal, rest assured that those who made such reports only looked at the surface, for other foods and other influences were having their effects on the system.

Many people die of food-poisoning and apoplexy. These bad results are due to wrong eating covering a long period and it is folly to blame the last meal. It would be queer if fruit and milk were not occasionally a part of the last meal. In winter, figs, dates or raisins with milk make an excellent lunch or breakfast. These fruits take the place of bread, for though they are not starchy, they contain an abundance of fruit sugar, which is more easily digested than the starch. Starch must be converted into sugar before the system can use it. On hot days milk and acid fruit make a satisfying meal. Many believe that milk and acid fruit should not be taken in the same meal, because the acid curdles the milk.

As we have already seen, the milk must be curdled before it can be digested. If this step in digestion is performed by the acid in the fruit no more harm is done than when it is performed by the lactic acid bacteria. Fruit juices and milk do not combine to form deadly poisons. If fruit and milk are eaten in moderation and no other food is taken at that meal the results are good. However, if fruit, milk, bread, meat, cake and pickles make up the meal, the results may be bad. Such eating is very common. But do not blame the fruit and the milk when the whole meal is wrong.

Likewise, if a hearty meal has been eaten and before this has had time to digest a lunch is made of fruit and milk, trouble may ensue. All the foods may be good, but a time must come when the body will object to being overfed. In summertime much less food is needed than during the cold months. Nevertheless, barring the Christmas holidays and Thanksgiving, people overeat more in summer than at any other time of the year. Picnics often degenerate into stuffing matches. We should expect many cases of serious illness to follow them, and such is the case.

Sometimes the milk is so carelessly handled that it becomes poisonous and at other times the fruit is tainted, but generally bad combinations and overeating are the factors that cause trouble when the fruit and milk combination is blamed. Buttermilk and clabbered milk are more easily digested by many than is the fresh milk. In Europe sour milk is a more common food than in this country. Here many do not know how excellent it is.

Two glasses of milk, or less, make a good warm-weather lunch. Those who have a tendency to be bilious should use cream very sparingly. Bilious people always overeat, otherwise their livers would not be in rebellion. The fat, in the form of cream, arouses decided protest on the part of overburdened livers.

A theory has found its way into dietetic literature, sometimes disguised as a truth, to the effect that boiled or hot milk is absorbed directly into the bloodstream without being digested. This is contrary to everything we know about digestion and assimilation, and although it is a fine enough theory it does not work out in practice. We have seen bad results when nothing but a small amount of the hot milk was fed to patients with weak digestive power. Perhaps others have had better results. When the system demands a rest from food, nothing but water should be given. Boiled or natural milk is then as bad as any other food, and worse than most, for in the absence of digestive power it soon becomes a foul mass, swarming with billions of bacteria. The system is compelled to absorb some of the poisons given off by the micro-organisms and the results are disastrous.

Every food we take must be modified by our bodies before entering the circulation, and milk is no exception. When milk is allowed to stand for a while the sugar ferments, through the action of the lactic acid bacteria. The sugar is turned into lactic acid, which combines with the casein and when this process has continued for a certain length of time the result is clabbered milk or sour milk. The length of time varies with the temperature and the care given the milk. If milk remains sweet for a long time during warm weather, discharge the milkman and patronize one whose product sours more quickly, for milk that remains sweet has been subjected to treatment.

All kinds of preservative treatment cause deterioration. If extraordinary care is taken with the milk and it is kept at a temperature of about forty-two degrees Fahrenheit, it may remain sweet five or six weeks, provided it is not exposed to the air, but such care is at present not practicable in commercial dairies. The milk contains unorganized ferments which spoil it in time without exposure to bacterial influences. These ferments cause digestion or decay of the milk.

Fresh butter is a palatable form of fat, which digests easily. Like all other milk products, it must be kept clean and cold, or it will soon spoil. Butter absorbs other flavours quickly and should therefore not be placed near odourous substances. It is best unsalted and in Europe it is very commonly served thus. When people learn to demand unsalted butter they will get good butter, for no one can palm off oleomargarine or other imitations under the guise of fresh unsalted butter.

Unsalted butter must be fresh or it will be refused by the nose and the palate. Salt and other preservatives often conceal age and corruption of foods. Butter combines well with starches and vegetables, in fact, it can be used in moderation with any other food, when the body needs fat. Butter should not be used to cook starches or proteins in. Greasy cooking should be banished from our kitchens.

Milk is a complex food, highly organized, and therefore is easily injured or spoiled. The general rule is that the more complex a food is, the more easily it spoils. It is rather difficult at present to get wholesome milk enough to supply the people of our large cities. When it is boiled, the milk keeps longer, but boiled milk is spoiled milk. The fine flavour is lost, the casein, which is the principal protein of milk, is toughened, the milk, which is normally a living liquid, is killed, the chemical balance is lost, the organic salts being rendered partly inorganic. Milk that is unfit to eat without being boiled is not fit to eat afterwards, for the poisonous end products of bacterial life remain.

The milk is soured by the bacteria it contains. The lactic acid bacteria are harmless. When there is a lack of care and cleanliness, other bacteria get into the milk, and these are also harmless to people in good health, and most of them are not injurious to sick people. The bacteria do not cause disease, but when disease has been established, they offer their kindly offices as scavengers. Bacteria thrive in sick people, especially when they are fed when digestive power is lacking.

Boiling retards the souring of milk, but when fat and protein are boiled together the protein becomes hard to digest. Milk is rich in both fat and protein. Excessive heat turns the milk brown, the milk sugar being carameled. Babies do not thrive on boiled milk. They may look fat, but instead of having the desirable firmness of normal children, they are puffy. Children fed on denatured milk fall victims to diseases very easily, especially to diseases which are due to lack of organic salts, such as rickets and malnutrition.

Pasteurization of milk is very popular. This is objectionable for the same reasons that boiling is condemned, though not to the same extent. Pasteurization is heating the milk to about 140 to 150 degrees Fahrenheit. This kills many of the bacteria, but many escape and when the milk is cooled off they begin to multiply and flourish again. It is estimated that pasteurized milk contains one-fourth as many bacteria as natural milk. So nothing is gained, and the milk is partly devitalized.

The advocates of pasteurization give statistics showing that milk so treated has been instrumental in decreasing infant mortality. But please bear in mind that previously a great deal of milk unfit for consumption was fed to the babies. Those who pasteurize milk generally are careful enough to see that they get a good product in the first place. If we can't get good milk we can do without it, for it is not a necessary food, but we can get good milk if we make the effort. If the milk is filthy, boiling or pasteurizing does not remove the dirt. Gauthier says of pasteurization: "Sometimes it is heated up to 70 degrees with pressure of carbonic acid. But even in this case pasteurization does not destroy all germs, particularly those of tuberculosis, peptonizing bacteria of cowdung, and the dust of houses and streets, etc."

Even boiling does not kill the spores of bacteria unless it is continued until the milk is rendered entirely unfit for food. To kill these spores it is necessary to boil the milk several times. The spores are small round or oval bodies which form within the bacterial envelope when these micro-organisms are subjected to unfavourable conditions.

The spores resist heat and cold that would kill almost any other form of life. When conditions are favourable they develop into bacteria again. After heating, the cream does not rise so quickly nor does it separate so completely as it does in natural milk. This is due to the toughening of the casein in the milk. Heating partly disorganizes the delicately balanced salts contained in the milk.

The result is that they can not be utilized so easily and completely by the body, for the human organism demands its food in an organic state, that is, in the condition built up by vegetation or by animals.

We may consume iron filings and remain anemic, in fact, the effect the iron medication has is to ruin the teeth, digestive organs and other parts of the body as a consequence. But if we partake of such foods as apples, cabbage, lettuce and spinach, the necessary salt is taken into the blood. Heating milk also makes it constipating. True, normal people can take boiled milk without becoming constipated, but how many normal people are there? We are sorely enough afflicted in this way now. Let us have a supply of natural milk or go without it.

It is not my desire to convey the impression that it does any harm to scald or boil milk occasionally, but if done daily it does harm, especially to the young. Scalded milk has its proper place in dietetics. Occasionally we find a person who has persistent chronic diarrhaea. If he is in condition to eat anything, this annoying affliction is usually overcome in a reasonable time if the patient will take boiled or scalded milk in moderation three times a day, and nothing else except water. How are we to obtain good milk? We can do it by using common sense, care and cleanliness.

It is well to remember that there are bacteria in all ordinary milk, and that if the milk is from healthy cows and is kept clean and cold these bacteria are harmless. Most of them are the lactic acid bacteria, which change the milk sugar into acid. When the milk has attained a certain degree of acidity, the lactic acid bacteria are unable to thrive and the souring process is slowed up and finally stopped. Most of the other bacteria in milk perish when lactic acid is formed.

This is why stale sweet milk is often harmful, when the same kind of milk allowed to sour can be taken with impunity. If the milk is kept in a cold place the bacteria multiply slowly. If it is kept in a warm place they increase in numbers at a rate that is marvelous, and consequently the milk sours much sooner. Even if the milk is kept cold, bacterial growth will soon take place, but it will perhaps not be lactic acid bacteria. It may be a form that causes the milk to become ropy and slimy or one that gives it a bad odour.

Bacteria are like other forms of vegetation, such as grass, weeds, flowers and trees, in that some flourish best under one condition and others under dissimilar conditions, and they struggle one against the other for subsistence and existence. Like flowers there are thousands of different forms of bacteria and they vary according to their food and environment. Peculiar odours in milk generally come from certain kinds of food given to the cows, such as turnips; from bacterial action; or from flavours absorbed from other foods or from odours in the air. Milk should not be exposed to odourous substances, for it becomes tainted very quickly.

Sometimes yeast finds its way into milk and causes decomposition of the sugar with the formation of carbon dioxide and alcohol. A count of the bacteria in milk often serves a good purpose, for it shows whether it is good and has had proper care. The consumers have a right to demand milk low in bacteria, for if no preservatives have been used, that means clean milk. If we could live in our pristine state of beatific bliss, if such it was, we would not have to use milk after childhood is past, but our present condition demands the use of easily digested foods and to many milk is almost a necessity.

The milk in the udder of a healthy cow is almost surely free from bacteria, but the moment it is exposed to the air these little beings start to drop into the fluid. The bacterial standards given by various city health departments vary. Those who are mathematically inclined may find the following figures interesting: In some great cities they allow 500,000 bacteria to the cubic centimetre of milk.

A cubic centimetre contains about twenty-five drops. In other words, they allow 20,000 bacteria per drop. This may seem very lively milk, but these bacteria are so small that about 25,000 of them laid end to end measure only about an inch, and it would take 17,000,000,000,000 of them to weigh an ounce, according to estimates. These are the tiny vegetables we hear and read so much about, that we are warned against and fear so much.

Truly the pygmies are having their innings and making cowards of men. The bacteria multiply by the simple process of growing longer and splitting into two, fission, as it is called, and the process is so rapid that within an hour or two after being formed a bacterium may be raising a family of its own.

Some of the milk brought to the cities contains as many as 15,000,000 bacteria per cubic centimeter, that is, about 600,000 per drop. This milk is either very filthy or it has been poorly cared for and should not be given to babies and young children. The filthiest milk may contain several billion bacteria to the cubic centimetre. By using care milk containing but 100, or even fewer, bacteria per drop can be produced.

From the standpoint of cleanliness this is excellent milk. Of course, the dairyman who takes pride enough in his work to produce such milk will sell nothing but what is first-class, and if he has business acumen he can always get more than the market price for his product.

The talk about germs has been overdone, but no one can deny that the study of bacteriology has made people more careful about foods. The filthy dairies that were the rule a few years ago are slowly being replaced by dairies that are comfortable, well lighted and clean. Do not allow the germs to scare you, for if ordinary precautions are taken no more of them will be present than are necessary, and they are necessary.

They thrive best in filth, and they are dangerous only to those who live so that they have no resistance. Wholesome milk can be produced only by healthy animals. Bovine health can be secured by the same means as human health. The cows must be properly fed and housed. They must have both ventilation and light. They must not be unduly worried. If a nursing of an angry mother's milk is at times poisonous enough to kill a baby, you may be sure that the milk from an abused, irritated and angry cow is also injurious.

If the animals are kept comfortable and happy they will do the best producing, both in quality and quantity. It may sound far-fetched to some to advocate keeping animals happy in order to get them to produce much and give quality products, but it is good science and good sense. Happy cows give more and better milk than the mistreated ones. The singing hens are the best layers. Cows should have fresh green food all the year, and this can be obtained in winter time by using silage.

It is a mistake to give cows too much of concentrated foods, such as oil meals and grains. Cattle cannot long remain well on exclusive rations of too heating and stimulating foods. When fed improperly they soon fall prey to various diseases, such as rheumatism and tuberculosis. It is the same with other domestic animals.

The horse when overfed on grain develops stiff joints. The hogs that are compelled to live exclusively on concentrated, heating rations are liable to die of cholera. Young turkeys that have nothing but corn and wheat to eat die in great numbers from the disease known as blackhead. It is the same law running all through nature, applying to the high and to the low, that improper nourishment brings disease and death. When cattle roam wild, the green grasses are their principal source of food. Man should be careful not to deviate too much, for forced feeding is as harmful to animals as it is to man.

The following excellent recommendations for the care of milk are given by Dr. Charles E. North of the New York City Milk Commission:

- "No coolers, aerators, straining cloths or strainers should be used.
- "The hot milk should be taken to the creamery as soon as possible.
- "The night's milk should be placed in spring or iced water higher than the milk on the inside of the can. It should not be stirred, and the top of the can should be open a little way to permit ventilation.
- "The milking pails and cans will be sterilized and dried at the creamery, and should be carefully protected until they are used.
- "Brush the udder and wipe with a clean cloth; wash with clean water and dry with a clean towel.
- "Whitewash the cow stable at least twice yearly.
- "Feed no dusty feed until after milking.
- "Remove all manure from cow stable twice daily.
- "Keep barnyard clean and have manure pile at least 100 feet from the stable.
- "Have all stable floors of cement, properly drained.
- "Have abundant windows in cowstables to permit sunlight to reach the floor.
- "Arrange a proper system of ventilation.
- "Do not use milk from any cows suspected of gargot or of any udder inflammation. Such milk contains enormous numbers of bacteria.
- "Brush and groom cows from head to foot as horses are groomed.
- "Use no dusty bedding; wood shavings or sawdust give least dust.
- "Use an abundance of ice in water tank for cooling milk."

Perhaps some will take issue with the doctor on the first paragraph of his recommendation. If straining cloths are used they should be well rinsed in tepid water, washed and then boiled. However, if his recommendations are carried out in letter and spirit no straining is necessary. Herr Klingelhofer near Dusseldorf, Germany, runs a model dairy. The cows, stables, milkers, containers, in fact, all things connected with the dairy are scrupulously clean. The milkers do not even touch the milk stools, carrying them strapped to their backs. The milk is strained through sterilized cotton and cooled. The cows are six and seven years old and are milked for ten or twelve months and they are not bred during this time. The first part of the milk drawn from each teat is not used, for that part is not clean, containing dirt and bacteria. This milk is practically free from bacteria, for without adding preservatives it will remain sweet, for as long as thirteen days.

If ordinary milk fails to sour in two or three days it shows that it has been treated. According to the Country Gentleman, it will cost from one cent and a quarter to one cent and three-quarters extra per quart to produce clean milk. Healthy adults can take milk teeming with bacteria without harm, but for babies it is best to have very few or none in the milk. At Dusseldorf the babies used to die as they do here when fed unclean milk.

Herr Klingelhofer says that when fed on his product "sterben keine". This is submitted to those who advocate pasteurizing the milk. Denatured milk makes sickly babies. Clean natural milk makes healthy babies. The extra cost of less than two cents a quart is not prohibitive. Most fathers, no matter how poor, waste more than that daily on tobacco and alcoholics.

The extra cost would be more than saved in lessened doctor bills, to say nothing of funeral expenses. The recompense that comes from the satisfaction of having thriving, sturdy, healthy children can not be figured in dollars and cents. Dr. Robert Mond, of London, after investigating for years, has come to the cease that sterilized milk predisposes to tuberculosis, instead of preventing it. He believes that milk so treated is so inferior that he would not personally use it. That sterilized milk predisposes to tuberculosis, as well as to other diseases which can attack the body only when it is run down, is natural.

Any food that has been rendered inferior can not build the robust health that comes to those who live on natural food. Adults who use sterilized milk should counteract its bad effects by partaking liberally of fresh fruits and vegetables. If the milk is clean, put into clean containers by careful milkers and is then kept cold until delivered, it will reach the consumers in good condition. Do not let the fact that when you consume a glass of milk you are also engulfing some millions of bacteria bother you, for bacteria are necessary to our existence. If all the bacteria on earth should perish, it would also mean the end of the human race.

Today the progressive farmer is coming to the fore. He is a man who is justly proud of his work, so it will probably not be long before all city people who desire clean milk can get it. The milk cure consists in feeding sick people on nothing but milk for varying periods. Generally the patient is told to either take great quantities three or four times a day, or to take smaller quantities perhaps every half hour. The milk cure has no special virtue, except that it is a monotonous diet.

The body soon rebels if forced to subsist on an excessive amount of but one kind of food. The individual loses his desire for food and even becomes nauseated. If the advocates of the milk cure would prescribe milk in moderation, instead of in excess, they would have better success. The benefit derived from the milk cure comes from the simplicity, not from the milk. A grape cure, an orange cure or a bread and milk cure would be as beneficial. The milk cure is ancient. It was employed twenty-five centuries ago.

BUTTERMILK AND SOUR MILK PRODUCTS

Buttermilk is a by-product of the butter-making process. The taste can be more or less sour depending on the sourness of the cream or the milk, which is used for butter making, or on the degree to which it sours after churning. It is also possible to make a product like soured buttermilk using milk or skimmed milk, by inoculating it with sour milk and letting it ferment for one day.

For the fermented milk you will need: fresh (skimmed) milk, a heat source, a wooden spoon, fresh fermented milk or buttermilk or a starter culture, a saucepan with a thick bottom, and a thermometer. Heat the fresh (skimmed) milk to boiling point, stirring all the time.

Cool it down to 18–20°C, for instance in a large pan with cold water. Add 10–30 ml of sour milk or buttermilk or a starter culture per each litre of milk (1%). Leave for it 18-24 hours at room temperature (18–20°C); if the surrounding temperature is higher, fermentation time will be somewhat shorter. After this the sour milk is ready. Store it in a cold place (cool basement or a refrigerator), if you want to keep it for some days.

RABI

Rabi is sweetened, concentrated milk. During concentrating, sugar is added from time to time.

You will need:

- (Unboiled) milk
- A heat source
- A wide, shallow iron pan with a thick bottom
- A flat metal scoop
- Sugar
- Scales.

Add sugar to the milk during the heating process (maximum 300 g per litre of milk) and follow the same procedure as for koa. Lumps of sugar will often be found in the end product.

YOGHURT

Yoghurt is produced when milk is soured by certain lactic acid bacteria, which prefer growing temperatures far above room temperature: 37–45°C. The milk should first be heated to 85°C or higher. A high pasteurisation temperature (above 72°C) gives a better consistency (thickness) to the final product. After the milk has been soured, the resulting yoghurt can be used to make more fresh yoghurt by adding it to fresh milk.

Basic Recipe for Yoghurt

You will need:

- Fresh raw milk
- A heat source
- A saucepan
- A spoon
- A thermometer
- Cooling facility (*e.g.* a large pan with cold water)
- A cool place (refrigerator or cellar)
- Starter culture for yoghurt or some fresh yoghurt
- Thermos flask or a box covered with a blanket

Heat the milk to 85°C or higher and keep it at this temperature for 3 minutes. Cool the milk to 45°C. Add 30 ml (2-3 tablespoons) of fresh yoghurt to each litre of milk; the yoghurt should not be more than 2 days old. Instead of fresh yoghurt you can use a yoghurt starter culture. Mix the milk and the starter and leave it to ferment. The time required for the milk to turn sour depends on the temperature.

To give you an idea:

- At 40-45°C it takes about 3 to 6 hours
- At 35-37°C it takes about 20 to 15 hours
- At 30°C it takes about 24 hours

The ideal temperature to make pleasant-tasting yoghurt with a firm consistency is 40-45°C. It is not possible to produce yoghurt at temperatures below 30°C or above 50°C. The correct temperature can be maintained using an insulated box or a blanket. Yoghurt is ready for consumption once the incubation period is finished. If cooled, yoghurt can be kept for one week.

Using a Thermos Flask

Heat the milk to at least 85°C, then cool it to 45°C. Pour 90% of the milk into a thermos flask, which has been rinsed with hot water. Mix 1-2 tablespoons of fresh prepared yoghurt (or yoghurt culture) with the rest of the milk and add this to the thermos flask. Close the flask well and leave to stand for 3-6 hours. Remove the yoghurt from the thermos flask and store in a cool place. Yoghurt made of sheep milk is very firm and therefore not suitable for fermentation in a thermos flask.

Yoghurt made from Milk Powder

Make milk from milk powder according to the instructions on the package, but add 10 to 15% extra milk powder. Dissolve the milk powder in water, heat it to boiling point and let it cool down to 45°C. Stir in 1-3 tablespoons of fresh yoghurt or a yoghurt culture per litre of milk. Cover the saucepan and put it in a warm, insulated place. After 3-6 hours this (firm and concentrated) yoghurt should be ready for consumption.

Remarks

- It is best to use fresh milk to make yoghurt. Milk powder can also be used. Sterilised milk may give a thinner yoghurt than pasteurised milk.
- After incubation, cooling is desirable, preferably below 10°C, so that souring is stopped (this retains the pleasant taste) and the bacteria remain more viable, allowing the yoghurt to be used to inoculate milk again.
- Make sure the milk ferments as quickly as possible, preferably at 40-45°C rather than 30°C. Harmful bacteria have less opportunity to develop if the fermentation process goes faster.
- Thicker yoghurt can be produced by adding 2 to 3 tablespoons of milk powder to each litre of milk before heating it to 85°C.
- It is not advisable to use fruit yoghurt from a shop as a starter culture because it contains a lot of additives. Plain yoghurt from a shop can be used if it is not too old. Sterilised yoghurt is not suitable either, because the yoghurt bacteria have been killed by the sterilisation process.
- When using yoghurt from a carton or a pot as a starter, first remove the top layer and take the yoghurt from the centre to make fresh yoghurt. This is because the bacteria in the middle are probably the most diverse and active.
- Stir the product as little as possible before removing some, to avoid the extra risk of incorporating undesirable bacteria.

GHEE

To make ghee, you will need:

- Butter
- A heat source
- A pan
- A metal spoon.

Heat the butter until water and fat form separate layers; the fat will float on top.

There are two ways to remove the water:

1. It can be removed by further heating. The water present will evaporate.
2. It is possible to remove the layer of fat with a spoon. This fat should then be heated again. The scum, which will form, has to be skimmed off regularly, preferably with a skimmer. The colour of ghee can vary from almost white to dark brown. A rancid flavour is acceptable, but if it tastes burnt it should be discarded.

KOA

Koa or khoa is a type of concentrated milk.

You will need:

- Fresh (unboiled) whole or skimmed milk
- A heat source
- A shallow, clean, wide, iron pan with a thick, flat bottom
- A flat, clean metal utensil to stir with (*e.g.* a flat pancake spatula).

Fill the pan with the milk up to 30 to 50% of its capacity. Bring the milk to boiling, stirring continuously. The water will evaporate and after some time the milk will reach a certain viscosity, this means the milk will thicken. Take great care to scrape the sides of the pan during stirring.

Once the milk has reached a dough-like consistency and stirring does not prevent the mass from sticking to the side of the pan, you can remove it from the heat source and lower the temperature by at least 20°C. At this point the water content should have been reduced to about 40% of what it was. Take the lump of koa out of the pan, put it on a cold surface and flatten it. After cooling, the koa will be firm and can be cut into squares. It will have a sweet, nutty flavour. Koa can be kept only for about 2-5 days because of possible re-infection. The preparation of koa requires much time (a few hours) and fuel, on top of which 1 litre of milk produces only 0.4 litres of koa.

KEFIR

Kefir, like yoghurt, is a milk product first discovered by the nomadic tribes living in the cold areas of the Caucasus. When making kefir, acid, gas and some alcohol are produced. Like sour milk and buttermilk, it has a special aroma which is different from that of yoghurt. Kefir is made by using a 'yoghurt plant', which is actually a misleading name as it is not a plant and it has nothing to do with yoghurt. The 'yoghurt plant' is in fact a cauliflower-like lump of chalky crystals and micro-organisms consisting of yeasts and bacteria.

The yeasts produce alcohol and gas, while the bacteria convert the milk sugar into lactic acid. You may be able to obtain a piece of this from someone who regularly makes kefir. If not, get some dried kefir granules at the local market. You will need: fresh raw milk, a saucepan, a heat source, a thermometer, a clean pan, a glass bottle with wide neck, a bottle which can be closed tightly, kefir granules or a 'yoghurt plant', a sieve, cooling facilities, clean water, a teaspoon, and a cool place to store the product.

Boil the milk and put it into a bottle, which has been thoroughly cleaned and rinsed with hot water. Do not fill the bottle completely; let the milk cool down to 20°C (use a thermometer). Add one tablespoon of kefir granules soaked in water to each half litre of milk in the bottle and cover it loosely so that the gas produced can escape.

You can use a fresh yoghurt plant instead of the soaked granules. Keep the bottle at a temperature of 16 to 18°C. After 24 hours the milk will have become a little thicker, forming some froth: this is kefir. Sieve the kefir and use the (washed with clean water) granules, which remain in the sieve to make fresh kefir again. The kefir is now ready for consumption or can ripen for some days to get more flavour.

Ripening of the kefir

The kefir can be left to 'ripen', during which time further fermentation takes place. This is essential to create kefir's characteristic qualities. Pour the kefir into a well-cleaned bottle, which can be closed, or a bottle with a clip fastening and do not fill more than 3/4 because gas forms during ripening. Leave the bottle at about 15°C, but do not store it for more than 3 days.

Towards the end of the ripening process the whey separates and can be incorporated again by stirring or turning the bottle. The end product is a thick, creamy, frothy drink with a sour taste and smell and the produced carbon dioxide is perceivable. If the kefir is left to ripen for more than 3 days, the milk may curdle and the drink becomes too sour. The temperature and the time are important as they determine the flavour. The kefir can be kept in a refrigerator or a cellar for a few days.

Storage of the Kefir Granules

If kefir production is stopped for some time, the kefir granules can be dried as follows: Put them into a sieve and rinse well with clean water until all remaining milk is removed. Place them on a clean cloth and leave to dry in a clean place, but not in the sun, until the granules have shrivelled up. The granules can be kept in a closed bottle in a cool place for 12 to 18 months. The rinsed granules can also be kept in a freezer. Kefir granules can also be kept in a pot with water at 4°C, but they become inactive after 8 to 10 days.

Remarks

- There is a greater chance of failure with kefir than with yoghurt. The end product may not only taste bad, it can also be unhealthy. In order to produce good kefir, the appropriate hygiene should be applied. Care should also be taken not to work at too high temperatures.

- The shelf life of kefir is the same as for yoghurt, sour milk and buttermilk: its quality will rapidly decrease if it is stored too long. At 5°C, soured products can be kept for about ten days; at 10°C this is three days, and 20°C is too high.
- Kefir may be too sour or too yeasty (due to an incorrect balance between the activity of the bacteria and yeasts). A fishy or ammonia-like taste can be the result of a breakdown of proteins by undesirable bacteria, which enter due to insufficient hygiene.

4

Dairy Operations

Dairy operations significantly affect the quantity and quality of manure that may be delivered to the anaerobic digestion system. In addition to the number of milk and dry cows, the housing, transport, manure separation, and bedding systems used by the dairy establishes the amount of material that must be handled and the amount of energy produced.

HOUSING SYSTEM

Confined dairy animals may be housed in a variety of systems. Commonly used housing systems include free stalls, corrals with paved feed lanes, and open lot systems. Milk cows, dry cows and heifers may be housed in free stalls, corrals, and open-lots on the same dairy. The type of housing used determines the quantity of manure that can be economically collected.

Free Stall Barns

Free stalls are currently the most popular method for housing large dairy herds. Free stall housing provides a means for collecting essentially all of the manure.

Fig. Typical Free Stall Barn with Center Feed Lane

Corrals

Corral systems with paved feed lanes are also commonly used. The manure deposited in the feed lanes can be scraped or flushed daily. From 40 to 55-per cent of the excreted manure may be deposited and collected from the corral feed lane. The balance of the manure may be deposited in the milk barn (10 to 15 per cent) or the open lot (30 to 50%).

Typically the manure deposited in the open lot is removed two to three times a year. It may have little net energy value after being stored in the open lot over prolonged periods of time. For corral systems one must make a reasonable determination of the recoverable manure deposited in the feed lane, corral, and milk barn. Corral systems also use a considerable amount of bedding material during the winter months. The straw bedding is generally removed in the spring and placed on the fields prior to spring planting.

Milk Barn

Dairy cows are milked two to three times a day. The cows are moved from their stalls to the milk parlor holding area. The milk parlor and holding area are normally flushed with fresh water. From 10 to 15 per cent of the manure is deposited in the milk parlor. In addition to the manure that is flushed, the cows may be washed with a sprinkler system. Warm water that is produced by the refrigeration compressors, vacuum pumps, and milk cooling system may be used for drinking water, manure flushing or washing the cows. It has been estimated that 5 to 150 gallons of fresh water per milk cow is used in the milking center. More common values are 10 to 30 gallons of fresh water per milk cow. The quantity and quality of water discharge from the milk parlor must be accurately measured. In many cases, the waste deposited in the milk barn is processed in a separate waste management system.

Open Lot

Fig. Open Lot System

In open lot systems the manure is deposited on the ground and scraped into piles. The manure is removed infrequently (once or twice a year). A significant amount of manure degradation occurs resulting in greenhouse gas emissions. In many cases, the open lot degradation produces manure that has little or no net energy value.

TRANSPORT SYSTEM

The commonly used manure transport systems are flush, scrape, vacuum, and loader systems. In free stall barns the manure can be flushed, scraped, or vacuum collected.

Flush Systems

If a flush system is used the manure is substantially diluted. The quantity of water used in a flush system depends on the width, length, and slope of the flush isle. The feed isles are generally 14 feet wide while the back isles are generally 10 feet wide. The slope varies between one and two per cent. A flush system will generally reduce the concentration of manure from 12 1/2 per cent solids, "as excreted", to less than one per cent solids in the flush water. Flush systems are however more economical and less labour-intensive than scrape or vacuum systems.

Fig. Free Stall Flush System-Flushing Feed Lane

Scrape Systems

Scrape systems are simply systems that collect the manure by scraping it to a sump. Under normal weather conditions the scraped manure has approximately the same consistency as the "as excreted" manure. During the warm dry summer manure may be dewatered on the slab.

Front-end Loader

Front-end loaders are used to stack and remove corral bedding and manure.

Vacuum Systems

Vacuum systems collect "as excreted" manure with a vacuum truck. Generally, the trucks collect approximately 4000 gallons per load. The manure can be hauled to a disposal site rather than to an intermediate sump. Vacuum collection is a slow and tedious process. The advantage is that the collected manure is undiluted and approximately equal to the "as excreted" concentration.

BEDDING

The type of bedding used can significantly alter the characteristics of the manure being treated. Typically straw, wood chips, sand, or compost are used as bedding material. In some cases paper mixed with sawdust is used. Compost usually has some sand mixed with the organic constituents. If composting is carried out on dirt lots, a significant amount of sand and silt may be incorporated into the compost.

Since anaerobic digestion will not degrade the wood chips, sand, or silt, it is necessary to remove those constituents prior to, or during anaerobic digestion process. The quantity of non-degradable, organic and inorganic material can significantly impact the performance of the anaerobic digester. The quantity of bedding added to the manure is a function of the design and operation of the dairy. Generally only the "kick-out" from the stalls is added to the manure. The quantity that is "kicked-out" is a function of the design of the dairy housing system as well as the type of bedding used.

MANURE PROCESSING SYSTEM

Each dairy has its own manure processing system. Scraped or flushed manure may be processed in a system separate from the milk barn waste, or the collected manure waste may be processed with the milk barn waste. In general, current manure processing consists of macerating the waste with a chopper pump, screening the waste to remove the organic fibres, followed by sedimentation to remove the sand, silt, and organic settable particles. Much of the degradable manure is removed during the separation processes. Up to 80% of the COD and 30% of the total Nitrogen and Phosphorous can be lost in the solids removed by the screen and sedimentation process. Detailed sampling and analysis is required to confirm losses.

HOLDING TANKS AND CHOPPER PUMPS

A wide variety of holding tanks and chopper pumps are used throughout the dairy industry. Typically, the tanks are relatively small but in some cases they are designed to hold several hours of flush water.

Primary Screens

An equally wide variety of screening systems are used. In many cases the screens are housed in separate enclosures to prevent freezing during the winter months. Outdoor screens are generally problematic during cold weather months. Fan separators are also used to provide efficient separation of the fibrous solids.

Fig. Primary Screens Background with Gravity Separators Foreground

The primary screens will remove a significant amount of degradable organic material that could be converted to gas in an anaerobic digester. The screened materials are generally used to produce bedding after being composted for the required time periods.

Gravity Separators

Gravity separators varying in size from 10 feet wide by 30' long to 24 feet wide by 80 feet long are usually placed after the primary screens. The purpose of the gravity separator is to remove the sands and silt present in the waste stream. If gravity separators are used without screening a thick mat of straw and fibres may develop on top of the gravity separator.

Fig. Floating Solids on Top of Gravity Separator

The gravity separators often incorporate weeping walls for the removal of liquid from the sedimentation chamber. The ability of the weeping wall to remove liquid waste depends on the periodic cleaning of the perforations to maintain flow. In many cases the gravity separators remove a significant amount of degradable organic material that could be utilized to produce gas. The COD test is a direct measure of the quantity of material that could be converted to methane gas. Recent tests have established that screen and gravity separators can remove 75% to 80% of the COD present in the waste stream. In one test the dairy parlor COD was reduced from 31,000 mg/l to 8,600 mg/l in the effluent from the gravity separator. In another the flush water influent to a separator system was 10,900 mg/L while the effluent was 1,800 mg/L. While a significant portion of the organic carbon (COD) is retained with the separated solids, an equal percentage of the nitrogen and phosphorus is not. The separation process alters the carbon to nitrogen ratio of both the liquid and solids streams.

Fig. Weeping Wall with Clogged Holes

The sedimentation process concentrates the organic solids, which are periodically removed. A recent analysis showed the flush water had a COD concentration of 25, 500 mg/l while the concentrated solids from the separator had a COD of 115, 800 mg/l.

Fig. Organic Solids on Top of Gravity Separator

Anaerobic decomposition of settled solids can be observed in separators that have a surface covered with methane gas bubbles. It is clear that existing solids handling practices contribute to greenhouse gas emissions and prevent efficient energy recovery from manure waste.

Fig. Gravity Separator with Anaerobic Decomposition of Organic Solid

EXPERIENCE IN DAIRY PRODUCT

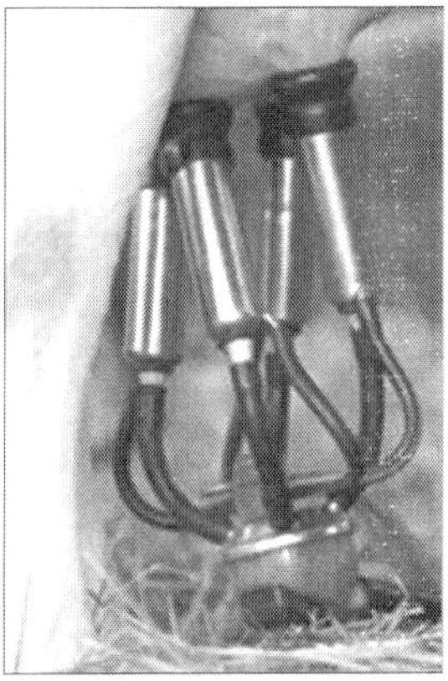

Milk is an emulsion containing about 87% water. The remaining 13% is made up of fatty acids, inorganic compounds, vitamins, and enzymes which do not dissolve in milk. Fresh milk is an ideal culture medium for microbial flora that can spoil the end product. Thus, the manufacture of milk and milk products is subject to very stringent rules.

These rules cover the way in which the livestock are kept and milked, preparation methods, additives, processing equipment, and the transport tanks

that move the milk from the farm to the processing plant. On its journey from the farm to the consumer, milk comes into contact with the walls of the equipment in which it is being processed or transported. The final product must reach the consumer meticulously clean and free from all germs. It is for this reason that the dairy industry has employed stainless steel for more than sixty years. The compact and smooth surface of stainless steel prevents bacterial contamination and facilitates cleaning and disinfection.

WHY STAINLESS STEEL

The development of stainless steel in the food industry is explained by the fact that it corresponds exactly to the requirements expected of materials in contact with food.

The requirements fall into one of the following three categories:

1. Chemical, bacteriological and organoleptic neutrality with regard to the food product.
2. Ability to be cleaned so the hygiene and appearance of the product is guaranteed.
3. Durability, including resistance to corrosion and aging.

Other factors also contribute towards making stainless steel the preferred material for the entire food processing industry. These include its mechanical characteristics, expansion coefficient, thermal conductivity and ease of use.

Chemical and Biological Neutrality

People are more aware than ever of the impact of the materials used to manufacture and transport food. Certain materials are widely distrusted because of problems with pollution, allergies and the volume of technical, scientific and medical information available. A number of studies have found that stainless has no adverse affects on human health, despite its wide and lengthy use in fields such as food, health and human hygiene. Stainless steel is a stable homogeneous alloy composed principally of iron, chromium (13 to 30%) and nickel (0 to 25%).

Cleaning and Disinfection

Cleaning involves the elimination of both visible and invisible contamination that adheres to the surface of a material. However, cleaning does not necessarily destroy the micro-organisms that are present on the material. Full bacteriological cleanliness is obtained only after surfaces are disinfected. Disinfection aims to reduce the microbial population on a surface to a level that is compatible with satisfactory hygiene practice and prevents contamination of the food.

Durability and Corrosion Resistance

Chromium, a component of stainless steel, forms a protective oxide compound on the surface of the metal. Known as a passive film, this layer slows and even prevents corrosion. If the surface of the metal is damaged, the passive film reconstitutes spontaneously at room temperature through the reaction of the chromium with water and oxygen. The stability of the passive layer is the determining factor that ensures stainless steels resist corrosion. Stability can be affected by the nature of the corrosive environment. This includes the oxidising power of the material that comes into contact with the stainless, its acidity level, chloride content and temperature. The corrosion resistance of stainless steel is very important when it is being cleaned or disinfected. When a piece of stainless steel equipment is being cleaned in place, only products that are highly alkaline or acid should be used to obtain perfect hygienic conditions. Stainless steels' exceptional resistance to corrosion of has enabled the dairy industry to develop widely and rapidly. Stainless behaves quite neutrally and does not alter the taste of fresh milk.

STAINLESS STEEL IN THE DAIRY INDUSTRY

The economic advantages and technical merits of stainless steel grades have been appreciated by the milk and dairy industry for a number of years. The following testimonials show that these benefits are becoming more widely understood.

CASE STUDY IN JAPAN

Dairy products go through a diverse range of processes as they are moved from the farms where the milk originates, to the production plants where it is turned into consumer products such as pasteurised milk, cream and yoghurt. Stainless steel is widely used in almost all of the processes the milk passes through on its journey from farm to consumer. In Japan, stainless steel products are used in all phases of the dairy production process.

These phases include:

- Transportation of the raw milk
- Inspection

- Purification
- Homogenisation
- Sterilisation
- Cooling
- Filling
- Final inspection
- Shipment to shops.

Different grades of stainless are used in each phase of dairy production. For example, grade SUS304 is utilised in products used to receive the milk and in all phases up to and including homogenisation. In the sterilisation and cooling processes, either grade SUS316 or SUS316L is used. During these processes, the products can reach temperatures as high as 130°C. These grades of stainless are also used in the production of cheese and butter, both of which contain salt. SUS316 or SUS316L can withstand both high tempera-tures and the corrosive properties of salt. Equipment used in ice-cream making is usually made from grade SUS304 as it can withstand the cold temperatures required.

Dairy products are very vulnerable to bacteria and have a short shelf-life. *To prevent deterioration of the product, all dairy machinery and equipment used in the production process should have the following characteristics:*

- *Product Integrity*: Equipment must have accurate control and measurement devices and there should be no leaks of oil or other contaminants from the equipment into the product.
- *Cleanliness*: Equipment must have smooth surfaces that do not allow bacteria to breed. They should also be no higher than the reach of an average person's hands to ensure they can be cleaned thoroughly.
- *Disassembly:* Equipment must be easy to take apart so it can be thoroughly cleaned and sterilised.
- *Inspection*: The design of the equipment should enable the visual inspection of the product during the production process.

In addition to hygiene, the characteristics of equipment used in the dairy industry include: resistance to chemical solutions used in cleaning (such as chloride-based sterilising agents, acids and alkalis); strength and workability.

Stainless steel meets all of these requirements. Other materials are often unsuitable because they cannot withstand the harsh conditions found in many food processing applications.

For example, sterilisers are subject to a lot of pressure. Alternative materials such as aluminium are not strong enough to withstand the pressure required. There is also a danger that aluminium will leach ions during cleaningin-place (CIP). Stainless can withstand high levels of pressure and does not leach during CIP.

TESTIMONIAL FROM SYNLAIT MILK LTD

The dairy industry is very dependent upon the use of stainless steel, for its excellent strength, its excellent corrosion resistance and the hygienic benefits of using a clean and green material. Also, the ability of the correct grade of stainless steel to be used in many dairy plant application-as long as the stainless steel plant is well fabricated and properly cleaned routinely.

The varied applications of stainless steel in a modern milk processing plant can be demonstrated as follows: Stainless steel is used to manufacture the truck-mounted milk tanker vessels that collect milk from the stainless steel tanks on dairy farms and deliver it to the dairy factory; The milk receiving plant at the dairy factory is made of stainless steel because of strict international sanitary and hygiene requirements; The milk and cream processing tanks at the dairy factory are all made of stainless steel; The large milk powder dryer vessels are made from stainless steel; All processing vessels for milk powder are stainless steels of various grades (usually types 304/316/2205); The milk powder bagging plant, the materials delivery lines are all in stainless steel.

MILK AND DAIRY PRODUCTS IN THE DIET

Diet refers to the way people feed themselves and the foodstuffs they use to do so. This is strongly influenced by people's traditions and religion, their economic position, their place in society and the possibilities offered by their natural surroundings. It is not surprising that each population group has its own diet. The use of milk and dairy products can also be looked at when examining the diet.

Here are a few examples of how the factors can influence the role and form of milk and dairy products in the diet:

- The cow is a sacred animal in India; therefore the rennet used in cheese making may not be taken from a calf's stomach.
- If milk or dairy products have to be bought, money is needed.
- In densely populated areas, people are forced to use all available land for crops that give a maximum yield, or crops that can be directly consumed by the people. This limits the land available for dairy farming.
- It may not be possible to keep cattle in certain regions, *e.g.* the humid tropics, due to the natural environment. For example, in humid areas of West Africa cattle cannot be kept because they are the host of the tsetse fly, which transmits sleeping sickness.

For these reasons, milk and dairy products in the diet can be of greater or lesser importance in one area or another. Economic and social situations are especially subject to change and dietary patterns change with them. New foodstuffs may be introduced to (partly) substitute others. Adoption of new products is often no easy matter; sometimes centuries-old traditions may have to change. Also, taste and other characteristics such as texture are important in the acceptance of new kinds of food. Something that may be considered very tasty in one place, may not be appreciated elsewhere.

Fig. Camel Milk is a very Popular Food in Parts of Africa and the Middle East

HYGIENE

Milk should be handled with care. There are several factors that can make milk go off and become unsuitable for further consumption.

These include:

- The presence of too many micro-organisms in the milk
- Contamination by diseased animals (tuberculosis, brucellosis) and/or people
- Bacterial and/or chemical conversion of certain substances in the milk
- Contamination of the milk with antibiotics (used for treatment of diseased animals), disinfectants, pesticides and so on.

The factors always cause some deterioration of milk. In some cases it is only the flavour that is affected, but usually the structure and smell of milk also change. In the case of contamination with antibiotics and disinfectants, the milk's appearance does not change, but fermentation, which is necessary for processing the milk, is inhibited. As suggested, first discuss the way micro-organisms cause the deterioration of milk. Then precautions which can be taken to minimise the impact of these factors will be explained, and some suggestions for cleaning and disinfection will be given.

DETERIORATION DUE TO MICRO-ORGANISMS

Bacteria, yeasts and moulds are all called: micro-organisms. Micro-organisms are very small and cannot be seen with the naked eye. They are found everywhere in nature: in the air, water, and soil and also in food and milk. Micro-organisms can multiply very rapidly. Milk in the udder of a healthy animal contains almost no microorganisms (aside from lactic acid bacteria). After the milk leaves the udder contamination with, sometimes harmful, micro-organisms will take place during milking, milk handling, transport and storage.

a. Shows badly cleaned utensils contaminating milk;
b. Improper milking also a source of contamination;
c. Shows multiplication of micro-organisms during storage;
d. Heating kills the micro-organisms

Milk may be contaminated with micro-organisms originating from the skin of the animal, the milker's hands, the milking utensils or the air. Most micro-organisms are not harmful, but some can cause diseases like salmonella infection, dysentery, tuberculosis (in man and in animals), diphtheria and typhoid. These disease-causing microorganisms are called pathogenic bacteria. Through inadequate hygiene, diseases can be transmitted from person or animal to person.

Micro-organisms can multiply very rapidly in milk. Temperature plays an important role in the life of micro-organisms. Their growth can start at a temperature of about 4°C. It is therefore very important to store milk or milk products at a temperature no higher than 4°C; otherwise deterioration will take place rapidly. Above 20°C, bacteria multiply at an incredible speed.

Most micro-organisms are killed during pasteurisation, *e.g.* at a temperature above 63°C for a period of at least 30 minutes. But a few, the so-called spore-forming bacteria, will survive more intense heating. They can give problems like off flavours and coagulation in pasteurised milk.

Yeasts and Moulds

Yeasts are micro-organisms that can ferment sugars into alcohol, gas and other substances. They are about 5-10 times larger than bacteria. Reproduction usually takes place through budding. Yeasts usually grow in an acid environment; they need oxygen and they can withstand rather high concentrations of acids. In dairy products, yeasts are usually found in soured products like sour milk or buttermilk, sour whey, butter, and curd and on the surface of cheese.

When present in large numbers, they produce gas and they cause undesirable off flavours of the product. Moulds are string-like micro-organisms. The fine threads, called mycelium, are large enough to be seen with the naked eye. To develop they need atmospheric oxygen, and they thrive best in humid and acid conditions. Moulds multiply by forming spores. These float easily through the air and can often be found on poorly maintained ceilings and walls. Their mobility makes them an important source of infection.

Moulds can be seen on the surface of butter or cheese in the form of coloured spots. For some soft cheeses (like Camembert and Brie) moulds are essential for ripening. In general, moulds are harmless, but some produce poisonous toxins (mycotoxins), such as aflatoxin in peanut products. Cells and spores of moulds and yeasts are destroyed by pasteurisation (heating the milk 30 minutes at 63°C or 20 seconds at 72°C).

Bacteria

Bacteria are single-cell micro-organisms that multiply by cell division. Raw milk and many dairy products contain many different kinds of bacteria. Environmental conditions (such as acidity, temperature, humidity or amount of oxygen) can change, making conditions less attractive for one group of bacteria but at the same time creating optimal conditions for another type. This is why some families of bacteria will always be found in milk or dairy products (lactic acid bacteria).

An exception must be noted, which is dried products like milk powder. Micro-organisms cannot grow without water and therefore the number of bacteria in uncontaminated milk powder will be low. Bacteria found in milk can be divided into two groups: useful and harmful. Lactic acid bacteria (*e.g.* Streptococcus lactis) are useful. They produce lactic acid, which is not harmful and gives milk a fresh, sour taste. Moreover lactic acid is a good preservative for the sour products.

Pathogenic bacteria (those that cause diseases in humans) cannot grow in acid products. When producing certain dairy products like soured milk, yoghurt and cheese, good use is made of these specific properties. Sometimes milk is spoiled by the growth of bacteria that do not produce lactic acid. In this case, certain disease-causing bacteria can develop and whey separates from the milk. This usually happens after long storage of pasteurised milk. The smell is unpleasant and the taste bitter. Such milk should not be consumed.

CONTAMINATION OF MILK WITH EXTRANEOUS MATTER

Extraneous substances must be prevented from entering into the milk. These can be dangerous to one's health or cause unpleasant flavours and smells, reducing the suitability of milk for further processing. Some examples are cleaning and disinfecting agents, medicine, pesticides and pieces of metal or glass. The feed given to animals, such as some weeds, onions and cabbage, can also influence the taste of the milk. This can be avoided by feeding the animals after milking. Good hygiene can reduce deterioration.

HYGIENIC PRODUCTION, STORAGE AND PROCESSING OF MILK

Contamination occurs when micro-organisms enter into the milk.

Possible sources of contamination during production, storage and processing are:

- Inflammation of the udder (mastitis)
- The animal itself: skin of teats and udder
- Conditions at the milking place (floor, dung, dust, dirty water, etc.)
- The person milking
- Utensils and equipment used during processing
- The air and environment

It is no easy task to keep micro-organisms out of milk. Much depends on the person who is milking, the care taken of the animals and the cleanliness of the utensils. If everything is well sanitised and kept clean, relatively few micro-organisms will enter into the milk. Good hygiene is of major importance. In addition, milk-if not used or processed immediately-should be cooled after milking and kept cool.

Good hygiene measures therefore:

- Prevent contamination of the milk.
- Prevent bacterial growth through good refrigeration of the milk.

Hygiene during Milking

There are several possible causes of contamination during milking. In a normal, healthy cow very low numbers of bacteria are found inside the udder and the teats. Cows possess various mechanisms to prevent the entry of bacteria. To avoid problems while milking, it is important that an animal become accustomed to the activity. It will then know that it will be milked, and will react positively to it. Such positive behaviour can start if, for instance, it hears milk cans clanging, feels its udder being cleaned, etc. Then the animal is easier to milk and gives more milk.

Stress and unrest make the cows move too much and kick; consequently more dirt and manure can enter into the milk. When a cow has an udder infection (mastitis), its milk will be contaminated with the bacteria that cause the udder infection, and that may produce pus and sometimes blood. Milk from these animals should not be used in any way. Mastitis can be prevented by maintaining good hygiene and avoiding injury to the teats during milking.

An infected udder is not always easy to see. When an udder infection occurs, it is advisable to remove milk from the udder very frequently (*e.g.* every 3 hours by hand). The number of microorganisms in the udder is thus reduced. Be aware, however, that milking an infected udder by machine or by hand is often painful for the animal. The animal will kick frequently and this can be an important source of contamination of healthy cows. Bacteria can be transferred from the skin or teats to the milk, even with healthy dairy cattle.

It is therefore important to clean the udder before milking. Wipe the udder clean with a dry, clean, preferably disposable cloth to prevent infection. If the teats or udder are really dirty, they must first be washed with clean, hand-warm water and a clean cloth and then dried with a clean towel. Cleaning the udder improves the cleanliness of the milk and makes milking easier. Skin and hair can also be sources of infection. Do not feed animals first before milking, it may create a lot of dust. The floor is clean, and be careful when clearing dung, mud or dust.

A clean, well-illuminated milking place and fresh surrounding air are essential to maintaining good hygiene. Insects such as flies and cockroaches can also be sources of infection. Try to control them as they can carry many bacteria and viruses. When milking, the milk is caught in a pail or bucket. Dirty milking equipment is the main source of infection of milk. If residues of milk remain in the equipment because of improper cleaning and drying, bacteria will develop in these residues. These bacteria are already accustomed to the milk and will multiply rather quickly during transport and storage of milk in the equipment.

Use pails and buckets that are smooth on the inside, for instance seamless metal buckets. All milking equipment should be thoroughly cleaned immediately after each use. Use soap or other detergents if necessary. Make sure that the water used is clean. If you are in doubt, boil it for several minutes or add chlorine. Very important: after cleaning, the equipment should be stored upside down in such a way that the inside of the buckets and cans dry. This prevents the remaining bacteria from growing.

The person milking plays the most important part in maintaining proper hygiene during production. He or she keeps an eye on the condition of the animal, chooses the milking place and cleans all the equipment. He or she should have clean hands and wear clean clothes. If the milker suffers from tuberculosis, salmonella infection, dysentery or some other disease, the risk of contamination of the milk becomes very high; it would be wise to have somebody else take over. This is also the case if the milker has open wounds or ulcers.

Hygiene during Storage and Processing

By now you should know that milk should be processed as quickly as possible after milking and that it should be properly stored in order to minimise its chances of spoiling. It is best to filter fresh milk through a filter or clean cloth. This will remove visible dirt that might have entered into the milk. Clean or replace the cloth during filtering or filter the milk several times. The cloth should be thoroughly cleaned after use and then left to dry in the sun.

In tropical conditions, raw milk, *i.e.* non-pasteurised milk, goes off within a few hours. It must therefore be kept cool and quickly pasteurised and again cooled to a temperature of 4°C if possible. Properly pasteurised and cooled milk can be kept for a few days, even in a warm climate. If you are not able to cool milk below 10°C, then do not mix different batches.

Even if the older milk is still good, you will end up with an increase in bacterial growth and reduction of the overall quality. Use clean equipment for storage. Containers that are clear, such as glass, should be stored in the dark as light reduces the quality of milk. Clean your equipment with clean water.

Cleaning and Disinfection

Utensils must be cleaned in such a way that all dirt, food residues, feed and micro-organisms are removed from the surface of the equipment. Dirty saucepans, jugs, milking equipment and utensils should be cleaned immediately after use. Washing soda (sodium carbonate) dissolved in hot water is an excellent cleaning agent. It may be useful to disinfect equipment in order to kill any remaining harmful micro-organisms. You can use a chloride solution such as bleach (sodium hypochlorite).

A proper way of cleaning your equipment is the following:

- Start cleaning immediately after milking, so that milk residues will not dry and stick on the buckets and utensils.
- Rinse well with water.
- Scrub the tools in a hot soda solution (1.5 tablespoons of soda to 5 litres of water), using a small amount of water to dissolve the soda before adding it to the rest of the water.
- Rinse well with hot water.
- Buckets, tubs, etc., should be turned upside down on a rack during storage; the water can then drain and no dirt or dust can enter. Let the utensils dry to prevent bacterial growth.

Well-cleaned tools are nearly sterile, only a small part of the bacteria remains on the tools. If these tools dry during storage hardly any bacteria will be present. In that case disinfection is not necessary. Tools which are used for storage of pasteurised milk or for cheese making and which do not get a heat treatment together with the milk can be disinfected after cleaning or before use.

Proceed as follows:

- Clean all your equipment properly. The following step will be ineffective if the utensils are not clean to start with.
- Disinfect in a chloride or bleach solution after cleaning or shortly before use. Add 2 tablespoons of bleach per 4. 5 litres of water.

It is advisable to use stainless steel equipment, cheesecloth and wooden utensils. Tools or any other equipment made from aluminium should not be washed in a strong soda solution, as soda attacks aluminium. Iron utensils will rust in a strong chloride solution. Therefore rinse and dry these utensils immediately after cleaning and disinfection.

If you have no cleaning agents-like soda-or disinfectants, you can disinfect your equipment as follows:

- Thoroughly clean the utensils using clean water.
- Rinse with a soap solution.
- Dry the equipment on a rack in the sun upside down or rinse with boiling water.

MILKING OPERATION

Milking machines are held in place automatically by a vacuum system that draws the ambient air pressure down from 15 to 21 pounds per square inch of vacuum. The vacuum is also used to lift milk vertically through small diametre hoses, into the receiving can. A milk lift pump draws the milk from the receiving can through large diametre stainless steel piping, through the plate cooler, then into a refrigerated bulk tank. Milk is extracted from the cow's udder by flexible rubber sheaths known as liners or inflations that are surrounded by a rigid air chamber. A pulsating flow of ambient air and vacuum is applied to the inflation's air chamber during the milking process.

When ambient air is allowed to enter the chamber, the vacuum inside the inflation causes the inflation to collapse around the cow's teat, squeezing the milk out of teat in a similar fashion as a baby calf's mouth massaging the teat. When the vacuum is reapplied in the chamber the flexible rubber inflation relaxes and opens up, preparing for the next squeezing cycle.

It takes the average cow three to five minutes to give her milk. Some cows are faster or slower. Slow-milking cows may take up to fifteen minutes to let down all their milk. Milking speed is only minorly related to the quantity of milk the cow produces — milking speed is a separate factor from milk quantity; milk quantity is not determinative of milking speed. Because most milkers milk cattle in groups, the milker can only process a group of cows at the speed of the slowest-milking cow. For this reason, many farmers will cull slow-milking cows.

The extracted milk passes through a strainer and plate heat exchangers before entering the tank, where it can be stored safely for a few days at approximately 42 °F. At pre-arranged times, a milk truck arrives and pumps the milk from the tank for transport to a dairy factory where it will be pasteurized and processed into many products.

AUTOMATIC MILKING MACHINES

They have been the topic of discussion for some time now, but as they become increasingly popular, Charlotte Johnston, TheCattleSite junior editor, looks at just how effective automatic milking machines are and evaluates their benefits.

A common myth regarding automatic milking machines is that they disengage the herdsman from the animals, and so neglect herd health. If this were true, the new Lely Astronaut A3 Next would not have won the 2009 Royal Association of British Dairy Farmers' Livestock and

Machinery Award at this year's Dairy Event. The award was presented on the grounds of economics, the welfare and close management of milking animals, the reduction of drudgery for family and employed staff, flexibility since it applies to family run and larger units, and the longer-term needs of the dairy industry.

TheCattleSite spoke to four dairy farmers around the world to find out how automatic milking equipment changed their lives. Click on *Tim Gibson*, *Doug Heintz*, *John Wolf* and *Max Warren* to read about their experiences.

HOW DOES IT WORK?

Very simply, a laser control will identify the location of the teats and a robust robot arm will attach cups individually to each teat from underneath the cow. Swift teat cup attachment is guaranteed even when milking cows with teats missing. Should a cup be kicked off, it is reconnected quickly.

COW ID SYSTEM AND COMPUTER MANAGEMENT SOFTWARE

Each cow has an electronic tag, which is read when she enters the machine. This allows the computer system to identify each cow. Management through the computer tells the stockman specific data for each individual cow.

The MQC sensor system checks the milk colour, conductivity and flow to ensure maximum control of milk quality. Comparing individual cow data to the average allows problems to be identified.

Milk identified with blood is separated and either discarded or diverted into another holding place. Furthermore, if a cow is being treated with medicine, it is possible for her to remain in the milking herd. This does require strict attention to detail - as accurate information must be input into the computer system as to what the cow has been treated with, how long the withdrawal period is etc.

When the cow comes in to be milked next, her milk will be automatically diverted into a separate holding area or disposed of. The same can be done for colostrum, for example, when a cow has recently calved as it may be beneficial to retain the colostrum for a couple of days. Entering all information on the system will allow this milk to be stored separately and after two days, her milk will return to the bulk tank. By checking the computer twice a day, the stockman can identify cows that are overdue milking or can treat illnesses identified immediately and the accessibility to such detailed data can allow accurate management decisions to be made.

DYNAMIC FEEDING

Dynamic feeding looks at an individual cow's feed requirements in relation to her yield. The programme takes into account milk yield, the cost of feed and the milk price received to work out the most economical feed rate for each individual cow. For Max Warren in Australia, this is particularly useful in a drought situation, where extra feed may have to be bought, but when milk prices are particularly low.

On average in the UK, for every litre of milk produced per cow, 0.3kg of dry matter is consumed, says Tim Gibson. This is standard regardless of whether they are high yielders. He believes he has saved one kilo of dry matter per cow per day, which with 140 cows equates to more than 51 tonnes of dry matter a year – in other words, a lot of money!

John Wolf and Doug Heintz from the US say that they now have the ability to feed a higher concentrate diet to higher yielding cows and they believe they have reduced feed costs.

The dynamic feeding programme allows optimum energy conversion efficiency. This unique and efficient system would not be possible without automatic milking machines.

LABOUR SAVINGS

All four farmers have cut back labour units. In Australia, Mr Warren believes he would require double the number of staff to milk 300 cows on a rotary system. Mr Wolf is saving 200 hours of labour a week, whilst Mr Gibson and Mr Heintz have cut back by almost one labour unit each. Adding up the cost of current labour, it is not hard to see how much money can be saved.

Lely estimate that for every one machine milking 70 cows, half a labour unit is saved. Two robots would milk 140 cows and save three-quarters of a labour unit, whilst three machines would milk 200 plus cows and save one full labour unit. "The biggest advantage by far is the labour saving. The flexibility has changed my life allowing me to spend more time with my family," says Mr Heintz.

BUILDING SPACE

Mr Gibson recommends that an area of approximately 15ft by 12ft is needed. The machines, however, are only 15ft by 6ft; the extra space takes into account a clean area behind the machine. With so little space required per machine, customers have the opportunity to increase indoor housing.

CONTROLLED GRAZING

Both Mr Wolf and Mr Heintz keep cows housed all year round; only dry cows and heifers may be put to grass. Mr Gibson allows his cows to graze outside during the summer with controlled grazing. Unlike the other farmers, Mr Warren's cows are grazed outdoors all year around on a three-way grazing system.

Mr Warren operated a two-way grazing system, where cows moved paddocks every 12 hours. He found with this system that cows were hanging around the parlour and so he changed the process to a three-way grazing system. Now, cows move fields every eight hours, each time passing through the parlour.

Cows are fed a wagon mix in the shed and fed concentrates when milked. All pasture is irrigated.

TRAINING HERD TO USE MACHINE

Mr Wolf said that it takes about three to four days to train heifers to use the machines. However, when the machines were introduced, it took about three weeks before things were running smoothly.

He believes that there will be about 0.5 per cent of each herd that does not adapt. Mr Heintz says that he spent three weeks pushing cows up to the machines, but it probably took about three months until the machines were working at full efficiency.

Over in Australia, Mr Warren says that it took two to three weeks to train the herd to walk into the robots but he felt it was probably twelve months until he was fully trained to use the system, the technology including computers was completely new to him!

HERD BENEFITS

The Lely Astronaut has a refined udder preparation procedure. Teats are cleaned individually and sprayed. The machine also automatically washes three times a day as well as washing in between each cow.

At the Robotic Dairy in Australia, cows are lasting six lactations. Further improvements in herd health can be seen by Mr Gibson reporting an increase in lifetime lactations per cow. Mr Wolf and Mr Warren reported improvements in foot health and reproductive performance.

Mr Heintz, who believes he had high standards of animal health prior to the installation of the machines, has reported no significant improvements in herd health but said that previous levels had been easily maintained. He has also only been using the machines for just under a year.

Walking around Mr Gibson's yard, it was apparent that the cows seemed dramatically calmer and at ease than in an conventional dairy unit. In fact, all farmers mentioned that cows were a lot less stressed, which in turn they believed led to better health and increased yields.

Increased yields seemed apparent across the board but the improvement varied between farms. In the UK, Mr Gibson reported an increase of 1,000 litres per cow over the last eight years, whereas Mr Heintz reported an increase of 10 to 15 pounds per day over the last ten months, which equates to 4562.5 pounds of milk per year or about an extra 2,000 litres per year.

SERVICE AND INSTALLATION

The same system operates in the US and the UK with Lely providing onsite support for the first 2-3 days of training cows. During the first two to three weeks, representatives will call in three or four times a week to help with the start-up.

Chad Huyser, Director of Sales and Operations for Lely US, says that farm management support staff will visit a couple of times during the first few months to do a thorough review of the farm management process, which includes making slight adjustments to feed rations and robot performance settings. Mr Gibson, who completed farm management training in Holland with Lely, says that he goes onto clients' farms to help customers adapt to the new style of management and prevent problems.

Lely also offer web training events with the farm management support staff, providing information to farmers on how to use the T4C, which is the onboard management software on the robots. Mr Huyser says that one of the areas that Lely focuses on globally is providing on-going support to all their installations.

Speaking with Mr Warren, he feels that ongoing technical support is essential, and is more than pleased with the service Lely continues to provide. He thoroughly enjoys working with the company and together, they have made a number of changes to adapt to Australian grazing conditions.

The new Lely Astronaut A3 robotic milking system requires servicing three times a year. Mr Gibson says the older models require servicing every eight weeks. He says that information from each service is sent back to a database at the headquarters, which allows engineers to access a vast amount of data to refer to in the case of a fault.

Due to the relative newness of the machines, there is no guarantee how long they will last. Mr Gibson believes that the oldest one in place is approximately 17 years old and still going strong. There have been no reports of machines wearing out to date. Lely estimates that products with the correct service and care will last at least 30 years. Machines can be constantly updated with new software, which comes at no extra cost to the customer.

Some people may be under the impression that robotic milking systems require someone to be on call 24/7 in case of a breakdown. This is not technically true. If there is an error in the system, an alarm system rings a telephone to tell whoever is on duty. During the day, this is not a problem but during the night, it could be more than a nuisance.

Not one of the farmers complained about this. During the evenings, Mr Gibson and Mr Warren said their machines may go off once a week. With seven machines, Mr Wolf is called three or four times a week. Mr Wolf stressed that the system can be manipulated and managed to reduce these calls through only five minutes of maintenance per robot per day.

HOW MUCH DOES IT COST?

The latest model, Lely Astronaut A3 Next, is around £110,000 for one robot. Opportunities for second-hand robots exist: all robots are refurbished at the Lely factory and come with a one-year warranty. Mr Gibson says that they often become available due to clients upgrading or retiring and cost approximately 75 per cent of the new price.

FARM BENEFITS

Mr Wolf explained that the introduction of robotic milking machines has allowed greater flexibility across the farm - particularly with regard to time schedules. "Having the robots in place means that I can now finish what I was doing in the field without rushing back to the parlour to milk the cows," he said. Mr Wolf also stressed that the working environment has significantly improved with cleaner areas for both cows and employees. But by far the biggest benefit for him was the decreased costs of production.

When asked about the biggest benefit of installing a robotic milking system, Mr Warren said the fact he was still farming. "Without robots, I would have stopped milking years ago. I really enjoy working with automatic milking machines and Lely. I love the technology that is always improving. It makes my work so much more enjoyable and interesting."

WHO SUPPLIES AUTOMATIC MILKING MACHINES?

If you are interested in robotic milking systems, Lely, BouMatic, Fulwood, DeLaval and GM Liberty are some of the suppliers.

MANAGEMENT TIPS

And finally, some tips from Mr Gibson, Mr Wolf, Mr Heintz and Mr Warren:

- Run your cows numbers and cash flow and make sure that the operation and introduction of robots into your herd would be viable
- Design the layout of the facility to how you want to manage it
- When designing the layout, consider if you may want to expand the enterprise and make sure this is possible
- Be prepared for a change in management
- Make sure you are prepared to get involved with the robots, checking data and ensuring careful attention to detail
- Be sure that whoever is supplying the machines will provide long-term full technical support, and
- See as many farms as possible and speak to as many people as possible before making any rash decisions!

Despite the success in Australia, to date there are still only six units operating automatic milking systems, which Mr Warren finds unbelievable. Despite the excellent support from Lely, Mr Warren is disappointed in the lack of support from the local dairy industry in Australia. Robots have started to take off in America and Canada, as well as across Europe. In the UK, Lely is running 250 robots on over 100 units. Worldwide there are 9,000 Lely robots. So if you fancy reducing the drudgery of milking or simply improving herd management, whilst increasing milk yields and herd health, perhaps it is time for you to open your mind to automatic milking systems?

MILK PROCESSORS

In order to estimate the potential success of these new product concepts, a survey instrument was sent to U.S. dairy processors. Participants were informed of the particulars of the new product concept and supplied with the estimated cost of manufacturing the new dairy product. Background information was requested in a variety of areas including the respondent's opinion as to why customers purchased their existing dairy products, the importance of private-label products, the size of the firm, and the frequency of the respondent's contact with end-customers.

The survey also asked whether the firm presently employed any membrane technology for dairy purposes (only 10 per cent did so), whether the respondent thought consumers would buy the new product, and requested suggestions for increasing the probability of the new product's commercial success. A total of 179 firms were contacted, of which 63 completed the survey for a total response rate of 35 per cent.

These 63 firms included 26 fluid milk processors, 21 ice cream manufacturers, and 16 yogurt manufacturers. The individuals surveyed were plant managers and/or those designated by each firm's receptionist as being most likely to make new product and/or new equipment decisions.

ANECDOTAL FINDINGS

Anecdotes reported by survey respondents are presented below. In all product areas, these comments illustrate a high level of cost and price sensitivity.

Based upon the results, one would expect this to be true for fluid milk. However, price sensitivity is also evident in the case of ice cream and even in the case of yogurt.

Comments received included:

- "Our current business isprice-driven, thus the supplier would have to absorb any cost increase." (From a bottler in the State of Washington.)
- "This product would have acceptance from only a small group of consumers." (From a bottler in Texas.)
- "We make a NFDM [Nonfat Dry Milk]-fortified product. At standard retail price it sells well. However, with a $0.05/gal. premium it does not sell well." (From a bottler in the State of New York.)
- "It takes a lot of consumer education and advertising to market a value-addedproduct at a premiumprice."(From a bottler in Kentucky.)
- "Adds cost. Our market area could not support this." (From an ice cream maker located in the North Central U.S.)
- "I am not sure customers would understand or care" (From an ice cream maker located in Wisconsin.)
- "This would take advertising." (From an ice cream maker in the High Plains.)
- "Customers currently love indulgence."(From an ice cream maker in California.)

- "Superiorflavour and texture at a competitive price will be a new product requirement." (From a yogurt maker in Ohio.)
- "I do not thinkyou can produce aproduct with enough improvement in taste/mouthfeel tojustify cost increase to the consumer." (From a yogurt maker in Michigan.)
- "Shelfplacement is critical." (From a yogurt maker in Illinois.)

STUDY IMPLICATIONS

These anecdotal findings reveal extreme concerns from dairy food manufacturers regarding highly elastic demand as well as the ability of consumers to perceive and/or pay for product differentiation. Because of cost, membrane technology investments for the purpose of producing a new product are likely to be scrutinized with great caution.

Among the three types of processors studied, statistical findings indicate that yogurt manufacturers are the most likely to consider a new functional-food formulation involving membrane technology. Concern about enhancing taste would likely be the motivation for making such a new product investment. This study has simply scratched the surface of an evolving industry process. Should membrane manufacturing technologies for new consumer dairy products become more widespread in the future, opportunities will then exist for economists to gain a greater understanding of this technology and its associated cost.

5

Sustainable Dairy Farming

Dairy farming is a class of agricultural, or an animal husbandry, enterprise, for long-term production of milk, usually from dairy cows but also from goats and sheep, which may be either processed on-site or transported to a dairy factory for processing and eventual retail sale. Most dairy farms sell the male calves born by their cows, usually for veal production, or breeding depending on quality of the bull calf, rather than raising non-milk-producing stock. Many dairy farms also grow their own feed, typically including corn, alfalfa, and hay. This is fed directly to the cows, or is stored as silage for use during the winter season. Additional dietary supplements are often added to the feed to increase quality milk production.

BACKDROP

Dairy farming has been part of agriculture for thousands of years. Historically it has been one part of small, diverse farms. In the last century or so larger farms doing only dairy production have emerged. Large scale dairy farming is only viable where either a large amount of milk is required for production of morc durablc dairy products such as chccsc, or thcre is a substantial market of people with cash to buy milk, but no cows of their own.

HAND MILKING

Centralized dairy farming as we understand it primarily developed around villages and cities, where residents were unable to have cows of their own due to a lack of grazing land. Near the town, farmers could make some extra money on the side by having additional animals and selling the milk in town. The dairy farmers would fill barrels with milk in the morning and bring it to market on a wagon. Until the late 1800s, the milking of the cow was done by hand. In the United States, several large dairy operations existed in some northeastern states and in the west, that involved as many as several hundred cows, but an individual milker could not be expected to milk more than a dozen cows a day. Smaller operations predominated. Milking took place indoors in a barn with the cattle tied by the neck with ropes or held in place by stanchions. Feeding could occur simultaneously with milking in the barn, although most dairy cattle were pastured during the day between milkings. Such examples of this method of dairy farming are difficult to locate, but some are preserved as a historic site for a glimpse into the days gone by. One such instance that is open for this is at Point Reyes National Seashore.

VACUUM BUCKET MILKING

The first milking machines were an extension of the traditional milking pail. The early milker device fit on top of a regular milk pail and sat on the floor under the cow. Following each cow being milked, the bucket would be dumped into a holding tank. This developed into the Surge hanging milker.

Prior to milking a cow, a large wide leather strap called a surcingle was put around the cow, across the cow's lower back. The milker device and collection tank hung underneath the cow from the strap. This innovation allowed the cow to move around naturally during the milking process rather than having to stand perfectly still over a bucket on the floor.

With the availability of electric power and suction milking machines, the production levels that were possible in stanchion barns increased but the scale of the operations continued to be limited by the labour intensive nature of the milking process. Attaching and removing milking machines involved repeated heavy lifting of the machinery and its contents several times per cow and the pouring of the milk into milk cans. As a result, it was rare to find single-farmer operations of more than 50 head of cattle.

STEP-SAVER MILK TRANSPORT

As herd size began to increase, the bucket milker system became laborious. A vacuum milk-transport system known as the Step-Saver was developed to transport milk to the storage tank. The system used a long vacuum hose coiled around a receiver cart, and connected to a vacuum-breaker device in the milkhouse, allowing farmers to milk many cows without the necessity of walking increasingly longer distances carrying heavy buckets of milk.

MILKING PIPELINE

The next innovation in automatic milking was the milk pipeline. This uses a permanent milk-return pipe and a second vacuum pipe that encircles the barn or milking parlor above the rows of cows, with quick-seal entry ports above each cow. By eliminating the need for the milk container, the milking device shrank in size and weight to the point where it could hang under the cow, held up only by the sucking force of the milker nipples on the cow's udder. The milk is pulled up into the milk-return pipe by the vacuum system, and then flows by gravity to the milkhouse vacuum-breaker that puts the milk in the storage tank.

The pipeline system greatly reduced the physical labour of milking since the farmer no longer needed to carry around huge heavy buckets of milk from each cow.

The pipeline allowed barn length to keep increasing and expanding, but after a point farmers started to milk the cows in large groups, filling the barn with one-half to one-third of the herd, milking the animals, and then emptying and refilling the barn. As herd sizes continued to increase, this evolved into the more efficient milking parlor.

MILKING PARLORS

Innovation in milking focused on mechanizing the milking parlor to maximize throughput of cows per operator which streamlined the milking process to permit cows to be milked as if on an assembly line, and to reduce physical stresses on the farmer by putting the cows on a platform slightly above the person milking the cows to eliminate having to constantly bend over. Many older and smaller farms still have tie-stall or stanchion barns, but worldwide a majority of commercial farms have parlors.

The milking parlor allowed a concentration of money into a small area, so that more technical monitoring and measuring equipment could be devoted to each milking station in the parlor. Rather than simply milking into a common pipeline for example, the parlor can be equipped with fixed measurement systems that monitor milk volume and record milking statistics for each animal. Tags on the animals allow the parlor system to automatically identify each animal as it enters the parlor.

RECESSED PARLORS

More modern farms use recessed parlors, where the milker stands in a recess such that his arms are at the level of the cow's udder. Recessed parlors can be herringbone, where the cows stand in two angled rows either side of the recess and the milker accesses the udder from the side, parallel, where the cows stand side-by-side and the milker accesses the udder from the rear or, more recently, rotary, where the cows are on a raised circular platform, facing the centre of the circle, and the platform rotates while the milker stands in one place and accesses the udder from the rear. There are many other styles of milking parlors which are less common.

HERRINGBONE AND PARALLEL PARLORS

In herringbone and parallel parlors, the milker generally milks one row at a time. The milker will move a row of cows from the holding yard into the milking parlor, and milk each cow in that row. Once all or most of the milking machines have been removed from the milked row, the milker releases the cows to their feed. A new group of cows is then loaded into the now vacant side and the process repeats until all cows are milked. Depending on the size of the milking parlor, which normally is the bottleneck, these rows of cows can range from four to sixty at a time.

ROTARY PARLORS

In rotary parlors, the cows are loaded one at a time onto the platform as it slowly rotates. The milker stands near the entry to the parlor and puts the cups on the cows as they move past. By the time the platform has completed almost a full rotation, another milker or a machine removes the cups and the cow steps backwards off the platform and then walks to its feed.

AUTOMATIC MILKER TAKE-OFF

It can be harmful to an animal for it to be over-milked past the point where the udder has stopped releasing milk. Consequently the milking process involves not just applying the milker, but also monitoring the process to determine when the animal has been milked out and the milker should be removed.

While parlor operations allowed a farmer to milk many more animals much more quickly, it also increased the number of animals to be monitored simultaneously by the farmer. The automatic take-off system was developed to remove the milker from the cow when the milk flow reaches a preset level, relieving the farmer of the duties of carefully watching over 20 or more animals being milked at the same time.

FULLY AUTOMATED ROBOTIC MILKING

In the 1980s and 1990s, robotic milking systems were developed and introduced. Thousands of these systems are now in routine operation. In these systems the cow has a high degree of autonomy to choose her time of milking within pre-defined windows. These systems are generally limited to intensively managed systems although research continues to match them to the requirements of grazing cattle and to develop sensors to detect animal health and fertility automatically.

CATTLE COWS

Dairy cattle (dairy cows) are cattle cows (adult females) bred for the ability to produce large quantities of milk, from which dairy products are made. Dairy cows generally are of the species Bos taurus. Historically, there was little distinction between dairy cattle and beef cattle, with the same stock often being used for both meat and milk production.

Today, dairy cows are specialized and most have been bred to produce large volumes of milk, with little or no regard for their production of meat. Between 1959 and 1990, US milk production doubled while the number of dairy cows declined 40 per cent.

COW

Dairy cows may be found either in herds on dairy farms where dairy farmers own, manage, care for, and collect milk from them, or on commercial farms. Dairy cow herds range in size from small farms of fewer than five cows to large herds of about 20,000. The average dairy farmer in the United States manages about one hundred cows but this varies from an average of 800 cows in California to under 80 in the North East states. Herd sizes vary around the world depending on landholding culture and social structure. In many European countries the average herd size is well below 50. In the UK it is over 100 in New Zealand 350 and Australia 280.

To maintain high milk production, a dairy cow must be bred and produce calves. Depending on market conditions, the cow may be bred with a "dairy bull" or a "beef bull." Female calves (heifers) with dairy breeding may be kept as replacement cows for the dairy herd. If a replacement cow turns out to be a substandard producer of milk, she then goes to market and can be killed for beef. Male calves can either be used later as a breeding bull or sold and used for veal or beef. Most dairy farmers begin breeding heifers at fifteen months of age. A cow's gestation period is approximately nine months, so most heifers give birth at around two years of age.

A Californian industry report, the natural lifespan of a dairy cow is approximately 15–25 years, however dairy cows are rarely kept longer than five years prior to slaughter. Herd life is strongly correlated with production levels. Approximately 17 per cent of the US beef supply comes from cull dairy cows: cows that can no longer be seen as a economic asset to the dairy farm. These animals may be sold due to common diseases of milk cows including mastitis, lameness, or other diseases. In India, the Hindu majority holds the cow as sacred and a motherly figure due to her capacity to give milk. Cow slaughter is legally banned in India (except in the state of Kerala, West Bengal and the seven north eastern states). Thus, spent dairy cows don't go to slaughter, but are often seen as roaming on the city streets and die of old age or disease. Some pious Hindu organizations manage "old age homes" for old dairy cows.

CALF

Market calves are generally sold at two weeks of age and bull calves may fetch a premium over heifers due to their size, either current or potential. Calves may be sold for veal, or for one of several types of beef production, depending on available local crops and markets. Such bull calves may be castrated if turnout onto pastures is envisaged, in order to render the animals less aggressive. Purebred bulls from elite cows may be put into progeny testing schemes to find out whether they might become superior sires for breeding. Such animals may become extremely valuable.

Most dairy farms separate calves from their mothers within a day of birth to reduce transmission of disease and simplify management of milking cows. Studies have been done allowing calves to remain with their mothers for 1, 4, 7 or 14 days after birth. Cows whose calves were removed longer than one day after birth showed increased searching, sniffing and vocalizations. However, calves allowed to remain with their mothers for longer periods showed weight gains at three times the rate of early removals as well as more searching behaviour and better social relationships with other calves.

After separation, most young dairy calves subsist on commercial milk replacer, a feed based on dried milk powder. Milk replacer is an economical alternative to feeding whole milk because it is cheaper, can be bought at varying fat and protein percentages. A day old calf consumes around 5 liters of milk per day.

BULL

A bull calf with high genetic potential may be reared for breeding purposes. It may be kept by a dairy farm as a herd bull, to provide natural breeding for the herd cows. A bull may service up to 50 or 60 cows during a breeding season. Any more and the sperm count will decline, leading to cows "returning to service" (to be bred again). A herd bull may only stay for one season since over two years old their temperament becomes too unpredictable. Bull calves intended for breeding commonly are bred on specialized dairy breeding farms, not production farms. These farms are the major source of stocks for artificial insemination (AI).

MILK PRODUCTION LEVELS

A cow will produce large amounts of milk over her lifetime. Certain breeds produce more milk than others; however, different breeds produce within a range of around 15,000 to 25,000 lbs of milk per lactation. The average for dairy cows in the US in 2005 was 19,576 pounds.

Production levels peak at around 40 to 60 days after calving. The cow is then bred. Production declines steadily afterwards, until, at about 305 days after calving, the cow is 'dried off', and milking ceases. About sixty days later, one year after the birth of her previous calf, a cow will calve again. High production cows are more difficult to breed at a one year interval. Many farms take the view that 13 or even 14 month cycles are more appropriate for this type of cow.

Dairy cows may continue to be economically productive for many lactations. Ten or more lactations are possible. The chances of problems arising which may lead to a cow being culled are however, high; the average herd life of US Holsteins is today fewer than 3 lactations. This requires more herd replacements to be reared or purchased.

Over 90 per cent of all cows are culled for 4 main reasons:

1. *Infertility*: Failure to conceive and reduced milk production. Cows are at their most fertile between 60 and 80 days after calving. Cows remaining "open" (not with calf) after this period become increasingly difficult to breed, which may be due to poor health. Failure to expel the afterbirth from a previous pregnancy, luteal cysts, or metritis, an infection of the uterus, are common causes of infertility.
2. *Mastitis*: Persistent and potentially fatal mammary gland infection, leading to high somatic cell counts and loss of production. Mastitis is recognized by a reddening and swelling of the infected quarter of the udder and the presence of whitish clots or pus in the milk. Treatment is possible with long-acting antibiotics but milk from such cows is not marketable until drug residues have left the cow's system.
3. *Lameness*: Persistent foot infection or leg problems causing infertility and loss of production. High feed levels of highly digestible carbohydrate cause acidic conditions in the cow's rumen. This leads to laminitis and subsequent lameness, leaving the cow vulnerable to other foot infections and problems which may be exacerbated by standing in feces or water soaked areas.
4. *Production*: Some animals fail to produce economic levels of milk to justify their feed costs. Production below 12 to 15 liters of milk per day are not economically viable.

Herd life is strongly correlated with production levels. Lower production cows live longer than high production cows, but may be less profitable. Cows no longer wanted for milk production are sent to slaughter. Their meat is of relatively low value and is generally used for processed meat.

REPRODUCTION

Since the 1950s, artificial insemination (AI) is used at most dairy farms; these farms may keep no bull. Advantages of using AI include its low cost and ease compared to maintaining a bull, ability to select from a large number of bulls to match the anticipated market for the resulting calves, and predictable results. More recently, embryo transfer has been used to enable the multiplication of progeny from elite cows. Such cows are given hormone treatments to produce multiple embryos. These are then 'flushed' from the cow's uterus. 7-12 embryos are consequently removed from these donor cows and transferred into other cows who serve as surrogate mothers. The result will be between 3 and 6 calves instead of the normal single, or rarely, twins.

HORMONE USE

Hormone treatments are given to dairy cows to increase reproduction and to increase milk production. The hormones are used to produce multiple embryos have to be administered at specific times to dairy cattle to induce ovulation. Frequently, for economic considerations, these drugs are also used to synchronize a group of cows to ovulate simultaneously. The hormones Prostaglandin, Gonadotropin Releasing Hormone, and Progesterone are used for this purpose and sold under the brand names Lutalyse, Cystorelin, Estrumate, Factrel, Prostamate, Fertagyl. Insynch, and Ovacyst. They may be administered by injection, insertion or mixed with feed.

About 17 per cent of dairy cows in the United States are injected with Bovine somatotropin, also called recombinant bovine somatotropin (rBST), recombinant bovine growth hormone (rBGH), or artificial growth hormone. The use of this hormone increases milk production from 11 per cent-25 per cent, but also increases the likelihood of cattle developing mastitis, reduction in fertility and lameness. The U.S. Food and Drug Administration (FDA) has ruled that rBST is harmless to people, although critics point out increased levels of insulin-like growth factor 1 (IGF-1) in milk produced using this hormone. The use of rBST is banned in Canada, parts of the European Union, Australia and New Zealand.

NUTRITION

Nutrition plays an important role in keeping cattle healthy and strong. Implementing an adequate nutrition programme can also improve milk production and reproductive performance. Nutrient requirements may not be the same depending on the animal's age and stage of production.

Forages, which refer especially to hay or straw, are the most common type of feed used. Cereal grains, as the main contributors of starch to diets, are important in meeting the energy needs of dairy cattle. Barley is one example of grain that is extensively used around the world. Barley is grown in temperate to sub-artic climates, and it transported to those areas lacking the necessary amounts of grain. Although variations may occur, in general, barley is an excellent source of balanced amounts of protein, energy, and fibre.

Ensuring adequate body fat reserves is essential for cattle to produce milk and also to keep reproductive efficiency. However, if cattle get excessively fat or too thin, they run the risk of developing metabolic problems. Scientists have found that a variety of fat supplements can benefit conception rates of lactating dairy cows. Some of these different fats include oleic acids, found in canola oil, animal tallow, and yellow grease; palmitic acid found in granular fats and dry fats; and linolenic acids which are found in cottonseed, safflower, sunflower, and soybean. It is also important to note that proper levels of fat also improve cattle longevity. Using by-products is one way of reducing the normally high feed costs. However, lack of knowledge of their nutritional and economic value limits their use. Although the reduction of costs may be significant, they have to be used carefully because animal may have negative reactions to radical changes in feeds. Such a change must then be made slowly and with the proper follow up.

PESTICIDE USE

A survey of the primary dairy producing areas in the US indicated that 13 per cent of lactating animals were treated with insecticides permethrin, pyrethrin, coumaphos, and dichlorvos primarily by daily or every-other-day coat sprays. Workers, particularly in stanchion barns, may be exposed to higher than recommended amounts of these pesticides.

BREEDS

In the United States, dairy cattle are divided into six major breeds. These are the: Holstein-Friesian, Brown Swiss, Guernsey, Ayrshire, Jersey, and Milking Shorthorn. In Rajasthan, an indigenous breed called Tharparkar exists, named from the Tharparkar District, now in Sindh Pakistan. Another type of dairy cow known as Nagauri from Nagaur District, the bull of which is renowned for its ability to plow fields and run. Traditionally, they used to pull covered wagons, known as rath, and in marriages to transport the newlywed couple. They are now a crutch for thriving agricultural and livestock rearing societies of the Thar Desert. Many other breeds are used nearly exclusively for beef, or for both dairy and beef purposes.

ANIMAL WASTE FROM LARGE DAIRIES

As measured in phosphorus, the waste output of 5,000 cows roughly equals a municipality of 70,000 people. In the U.S., dairy operations with more than 1,000 cows meet the EPA definition of a CAFO, and are subject to EPA regulations. For example, in the San Joaquin Valley of California a number of dairies have been established on a very large scale.

Each dairy consists of several modern milking parlor set-ups operated as a single enterprise. Each milking parlor is surrounded by a set of 3 or 4 loafing barns housing 1,500 or 2,000 cattle. Some of the larger dairies have planned 10 or more series of loafing barns and milking parlors in this arrangement, so that the total operation may include as many as 15,000 or 20,000 cows. The milking process for these dairies is similar to a smaller dairy with a single milking parlor but repeated several times.

The size and concentration of cattle creates major environmental issues associated with manure handling and disposal, which requires substantial areas of cropland for manure spreading and dispersion, or several-acre methane digesters. Air pollution from methane gas associated with manure management also is a major concern. As a result, proposals to develop dairies of this size can be controversial and provoke substantial opposition from environmentalists including the Sierra Club and local activists. The potential impact of large dairies was demonstrated when a massive manure spill occurred on a 5,000-cow dairy in Upstate New York, contaminating a 20-mile stretch of the Black River, and killing 375,000 fish. On Aug. 10, 2005, a manure storage lagoon collapsed releasing several million gallons of manure into the Black River. Subsequently the New York Department of Environmental Conservation mandated a settlement package of $2.2 million against the dairy.

USE OF HORMONES

It is possible to maintain higher milk production by injecting cows with growth hormones known as recombinant BST or rBGH, but this is controversial due to its effects on animal and possibly human health. The European Union, Japan, Australia, New Zealand and Canada have banned its use due to these concerns. In the US however, no such prohibition exists, and approximately 17.2 per cent of dairy cows are treated in this way. The U.S. Food and Drug Administration states that no "significant difference" has been found between milk from treated and non-treated cows but based on consumer concerns several milk purchasers and resellers have elected not to purchase milk produced with rBST.

MANAGEMENT OF THE HERD

Modern dairy farmers use milking machines and sophisticated plumbing systems to harvest and store the milk from the cows, which are usually milked two or three times daily. During the warm months, in the northern hemisphere, cows may be allowed to graze in their pastures, both day and night, and are brought into the barn only to be milked. Many barns also incorporate tunnel ventilation into the architecture of the barn structure.

This ventilation system is highly efficient and involves opening both ends of the structure allowing cool air to blow through the building. Farmers with this type of structure keep cows inside during the summer months to prevent sunburn and damage to udders. During the winter months, especially in northern climates, the cows may spend the majority of their time inside the barn, which is warmed by their collective body heat.

Even in winter, the heat produced by the cattle requires the barns to be ventilated for cooling purposes. Many modern facilities, and particularly those in tropical areas, keep all animals inside at all times to facilitate herd management. Housing the cow can be either loose housed or stalls. There is little research available on dimensions required for cow stalls, and much housing can be out of date, however increasingly companies are making farmers aware of the benefits, in terms of animal welfare, health and milk production. In the southern hemisphere cows spend most of their lives outside on pasture, although they may receive supplementation during periods of low pasture availabliity.

The production of milk requires that the cow be in lactation, which is a result of the cow having given birth to a calf. The cycle of insemination, pregnancy, parturition, and lactation, followed by a "dry" period of a few weeks before calving which allows udder tissue to regenerate. Dairy operations therefore include both the production of milk and the production of calves. Bull calves are either castrated and raised as steers for beef production or veal.

HEALTH AND WELL-BEING

Common ailments affecting dairy cows include infectious disease, metabolic disease and injuries caused by their environment.Lameness is commonly considered one of the most significant animal welfare issues for dairy cattle. It can be caused by a number of sources, including infections of the hoof tissue and physical damage causing bruising or lesions. While housing and management features common in modern conventional dairy farms have been identified as contributing risk factors, small farms in developing countries can also demonstrate high rates.

DAIRY PRODUCT

Dairy products are generally defined as foods produced from cow's or domestic buffalo's milk.

They are usually high-energy-yielding food products. A production plant for such processing is called a dairy or a dairy factory. Raw milk for processing mostly comes from cows and to a lesser amount from domestic buffalos, but occasionally from other mammals such as goats, sheep, yaks, or horses. Dairy products are commonly found in European, Middle Eastern and Indian cuisine, whereas they are almost unknown in East Asian cuisine.

TYPES OF DAIRY PRODUCTS

- Milk after optional homogenization, pasteurization, in several grades after standardization of the fat level, and possible addition of bacteria Streptococcus lactis and Leuconostoc citrovorum
  - Crème fraîche, slightly fermented cream
    a. Smetana, Central and Eastern European variety of sour cream
    b. Clotted cream, thick spoonable cream made by heating
  - Cultured buttermilk, fermented concentrated (water removed) milk using the same bacteria as sour cream
  - Kefir, fermented milk resembling buttermilk but based on different yeast and bacteria culture
  - Kumis/Airag, slightly fermented mares' milk popular in Central Asia
  - Milk powder (or powdered milk), produced by removing the water from milk

a. Whole milk products
b. Buttermilk products
c. Skim milk
d. Whey products
e. Ice Cream
f. High milk-fat and nutritional products (for infant formulas)
g. Cultured and confectionery products

- Condensed milk, milk which has been concentrated by evaporation, often with sugar added for longer life in an opened can
- Evaporated milk, (less concentrated than condensed) milk without added sugar
- Ricotta cheese, milk heated and reduced in volume, known in Indian cuisine as Khoa
- Infant formula, dried milk powder with specific additives for feeding human infants
- Baked milk, a variety of boiled milk that has been particularly popular in Russia

• Butter, mostly milk fat, produced by churning cream
  - Buttermilk, the liquid left over after producing butter from cream, often dried as livestock food
  - Ghee, clarified butter, by gentle heating of butter and removal of the solid matter
  - Smen, a fermented clarified butter used in Moroccan cooking.
  - Anhydrous milkfat
• Cheese, produced by coagulating milk, separating from whey and letting it ripen, generally with bacteria and sometimes also with certain molds
  - Curds, the soft curdled part of milk (or skim milk) used to make cheese (or casein)
  - Whey, the liquid drained from curds and used for further processing or as a livestock food
  - Cottage cheese
  - Quark
  - Cream cheese, produced by the addition of cream to milk and then curdled to form a rich curd or cheese made from skim milk with cream added to the curd
  - Fromage frais
• Casein
  - Caseinates
  - Milk protein concentrates and isonates

  - Whey protein concentrates and isonates
  - Hydrolysates
  - Mineral concentrates
- Yogurt, milk fermented by Streptococcus salivarius ssp. thermophilus and Lactobacillus delbrueckii ssp. bulgaricus sometimes with additional bacteria, such as Lactobacillus acidophilus
  - Ayran
  - Lassi
- Clabber (food), milk naturally fermented to a yogurt-like state
- Gelato, slowly frozen milk and water, lesser fat than ice cream
- Ice cream, slowly frozen cream and emulsifying additives
  - Ice milk
  - Frozen custard
  - Frozen yogurt, yogurt with emulsifiers that is frozen
- Other
  - Viili
  - Kajmak
  - Filmjölk
  - Piimä
  - Vla
  - Dulce de leche

HEALTH RISKS OF CONSUMING DAIRY PRODUCTS

Most dairy products contain large amounts of saturated fat. Some dairy products may cause health issues for individuals who have a lactose intolerance and milk allergies. Some dairy products such as blue cheese may become contaminated with the fungus Aspergillus fumigatus during ripening, which can trigger asthma and other respiratory problems in susceptible individuals. Vegans and some vegetarians avoid dairy products due to a variety of ethical, dietary, environmental, political, and religious concerns.

MEMBRANE TECHNOLOGY IN DAIRY PROCESSING

Membrane filtration technologies, such as ultra filtration and reverse osmosis, are capable of the molecular fractionation of fluids. Milk is ideally suited for processing by membrane filtration because it is a fluid consisting largely of water, lactose, butterfat, and protein molecules. Separation at the molecular level means that butterfat, lactose, and protein can be isolated from one other. Through the use of cellulose filters and high pressure pumps, membrane technologies take the two-dimensional concept of the venerable cream separator (*i.e.*, milk in, cream and skim out) into the third dimension and even beyond.

Membrane technologies have brought about substantial change in the dairy industry. However, because of the rapid pace of innovation many new dairy products created by membrane technology have not yet gained effective consumer demand. As David Hettinga, Vice President and Chief Technical Office of Land O'Lakes, stated, "one of the problems with this technology is we have a product or a technology chasing the market". The purpose of this research is to introduce readers to the membrane process and then attempt to assess the consumer demand for a few such new products.

The traditional dairy manufacturing paradigm has been to separate whole milk into cream and skim milk using a centrifuge. The skim milk is then John W. Siebert is an associate professor and Alejandro Lalor is a graduate research assistant, Department of Agricultural Economics, Texas A&M University, College Station, TX 77843-2124. Sung-Yong Kim is a research associate at the Korea Rural Economic Institute. The authors wish to thank the Southwest Dairy Farmers of Sulphur Springs, TX and Texas A&M University for their support of this research. The authors also with to thank two anonymous reviewers as well as Dr. Sefa Koseoglu. often evaporated to produced condensed skim.

Most dairy products are made using various combinations of milk, cream, skim, and condensed skim. The shortfall of this traditional technology approach is that protein and lactose (the main ingredients of skim) are bound to one another. A key value of membrane technology is that it enables a separation of these two ingredients. With membrane technology, protein, butterfat, and lactose can be used to manufacture dairy products more directly. Should sales of milk increase due to the development of new products, total dairy-farmer income would be likely to increase as well. With approximately 590,000,000 pounds of nonfat dry milk currently in government warehouses, research into demand expansion remains a high priority for dairy farmers.

TECHNOLOGY REVIEW

The most widely accepted dairy applications of membrane technology have been cost-reducing in nature. For example, most modern cheese plants use membrane technology to extract valuable protein isolates from the whey stream. Whey-protein concentrate is currently an important source of income to all large cheese makers. The portion of a modern cheese plant devoted to whey-product manufacturing and storage can be almost as large as that devoted to cheese. Due solely to the ability of membrane technology to extract protein from whey, whey is no longer a disposal problem-it is now a profit center.

In New Zealand, membrane technology is used to produce a powdered dairy product consisting largely of butterfat and protein. Due to its functionality, this ingredient-called dry ultrafiltered milk or milk protein concentrate (MPC)-can be used to make cheese. MPC is imported to the United States for the purpose of boosting cheese-plant yields. In this regard it is a substitute for domestic nonfat dry milk, and for this reason has been viewed as a threat to the U.S. milk price support programme.

Dairy farmers in remote regions of the United States have used membrane technology to reduce raw milk transportation costs. At the farm, ultrafiltration is being used to remove lactose and water from milk. Also at the farm, reverse osmosis is being used to remove water from milk. Membrane technology will likely replace the traditional cheese vat in the future. The traditional cheese vat is a large kettle (*e.g.*, 5,000-gallon capacity) that uses calf rennet, heat, and agitation in order to yield cheese curd and whey from milk.

The curds are then pressed into blocks of cheese and aged. Membrane technology has the potential to produce cheese by molecular separation of lactose from the butterfat and protein. This would allow the design of equipment that accepts milk as an input and produces liquid cheese and whey as outputs. The liquid cheese stream could then be poured into forms for hardening and aging.

The objective of this research was to determine if membrane technology has the potential to create new commercial dairy products for direct purchase by consumers. The key questions investigated concern the capabilities of membrane technology, the economics of producing new consumer products, and the consumer market potential of any such new products. Before new consumer dairy products such as these can be found and evaluated, the technology must first be understood. Figure provides a membrane technology diagram.

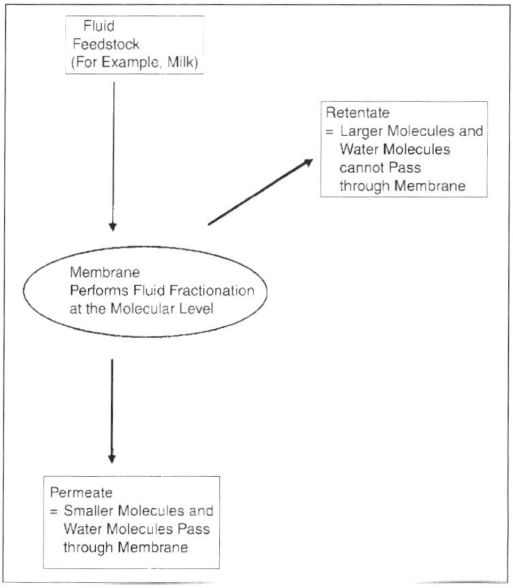

Fig. The Membrane Filtration Process is Initiated by a Fluid being Pumped over a Membrane. This Causes Fractionation at the Molecular Level.

This figure shows a fluid being pumped across a membrane under high pressure. Smaller particles pass through the membrane and are termed permeate. Larger particles cannot pass through and are denoted as retentate.

The membrane filtration process can be performed at progressive levels of molecular selectivity. Reverse Osmosis (RO) is a term denoting a very fine membrane-filtration process. To understand filtration in its application to dairy, consider that raw milk consists largely of water, lactose, butterfat, protein, and minerals.

Applying RO to milk would thus produce a permeate which consists mainly of water and a retentate which consists of water, lactose, butterfat, protein, and minerals. Ultrafiltration (UF) allows somewhat larger molecules to pass through the membrane than does RO.

In this case, not only water but also lactose molecules will pass through the membrane. Thus, applying UF to milk produces both a permeate consisting of water and lactose and a retentate consisting of water, lactose, butterfat, and protein.

EQUIPMENT INDUSTRY SURVEY

To gain an understanding of the current status of membrane technology in the dairy industry, the authors made an initial survey of the membraneequipment industry. Our objective was to learn about potential new consumer dairy product applications of membrane technology. We contacted nineteen firms involved in various combinations of equipment manufacturing, facilities and/or equipment design, and equipment installation. The authors found these nineteen firms through advertisements in the dairy trade press, through suppliers listed in the International Dairy Foods Association Membership Directory, and through attendees at a Texas A&M University Short Course on Membrane Technology.

The authors do not know what percentage of the dairy membrane manufacturing industry was contacted through their survey but the percentage is believed to be high, as all known firms were contacted. Also, the supplier industry is relatively concentrated. Thus, despite the small number of firms involved, this sample should be considered representative of the dairy membrane equipment industry during 1999. The firms contacted served the entire United States. Nine of the thirteen firms were headquartered in either Minnesota or Wisconsin. Several of the firms were subsidiaries of international companies.

Thirteen of the nineteen firms contacted participated, a response rate of 68 per cent. The responding firms viewed membrane technology as advancing rapidly in terms of fractionation selectivity, methods, and reliability. Technological advances usually originate in Australia, New Zealand, or Western Europe. Consequently, U.S. firms often employ technology after it has proven its value elsewhere. Two dairy industry forces, when taken in combination, likely explain why New Zealand, Europe and Australia have historically taken the lead in the development of membrane technologies.

First, the U.S. Food and Drug Administration has restricted dairy manufacturers from using membrane technology in the production of traditional dairy products such as cheese. The FDA must approve on a firm-by-firm and case-by-case basis that the product manufactured with membrane technology has no compositional or organoleptic differences when compared to a product made in full conformance with regulatory Standards of Identity.

Second, the U.S. pricesupport programme only offers stand-by purchasing authority for cheese, butter, and nonfat dry milk; therefore, membrane-based dairy products such as milk protein concentrate powder would not qualify for the programme.

As a result, U.S. dairy-industry investment is often made in traditional production technology in order to reduce exposure to price risk. Manufacturers were asked about the future of membrane technology. The consensus was that byproduct extraction at the dairy processing plant would be the main area for the future impact of membrane technology.

Specifically, eight of twelve manufacturers who responded to a question concerning whether the biggest impact of membrane technology would be at the processing plant or at the farm felt the biggest impact would be at the plant. Ten of eleven manufacturers who responded to a question concerning whether the biggest impact of membrane technology would be upon dairy products or dairy by-products (*e.g.*, on cheese as opposed to cheese whey) felt that the biggest impact would be in the by-product area.

THREE NEW PRODUCT CONCEPTS

Seven of ten manufacturers who responded to a question concerning whether membrane technology would be better at producing new dairy products or existing dairy products felt that the biggest impact would be upon new products.

The following new consumer product ideas were gleaned from the membrane industry:

- Protein-fortified, 2% reduced-fat milk can be made by a combination of whole milk, skim milk, and skim milk retentate. This product recipe had 18 per cent more protein than regular 2% reduced-fat milk without increased lactose levels.
- High-protein, low-lactose ice cream can be made by a combination of sweet cream, skim milk retentate, and nonfat dry milk. The desirability of this product results from substituting protein for lactose. The recipe evaluated had 48 per cent more protein and 32 per cent less lactose than regular ice cream.
- Nonfat yogurt can be made with more protein and therefore less stabilizers. This product can be made by a combination of skim milk, skim milk retentate, and nonfat dry milk. The particular product recipe

evaluated had 15 per cent more protein and 21 per cent less lactose than regular nonfat yogurt.

These products all substitute protein for lactose. The addition of protein can bring more product body, better mouthfeel, and higher product viscosity. The reduction of lactose brings little in the way of reduced sweetness as lactose has only onesixth to one-third the sweetness of sucrose. Any such loss of sweetness can easily be countered by the addition of a small amount of sugar.

Focusing just on yogurt, the major benefits are two-fold. First, less product separation will occur. In other words, less liquid whey will form and separate from the yogurt curd. Second, if yogurt is made using non-dairy stabilizers, then product label-purity is compromised.

The non-dairy stabilizers which might be used for this purpose could include any of the following ingredients:

- Starch,
- Pectin,
- Gelatin,
- Vegetable gums (carboxymethyl cellulose,
- Locust bean, or guar), or seaweed gums (such as alginates or carrageenans).

To understand the important role of protein and why increased protein content is beneficial, consider the properties of two well-known dairy products, cheese and butter. The major difference between cheese and butter is that butter contains 80 per cent milkfat, while cheddar cheese contains approximately 32 per cent milkfat and 31 per cent protein. Even though butter contains less moisture than cheddar cheese, it remains a softer product. In addition, butter's weak texture makes it unsuitable for eating out of hand while a substantial amount of cheese is eaten in this fashion.

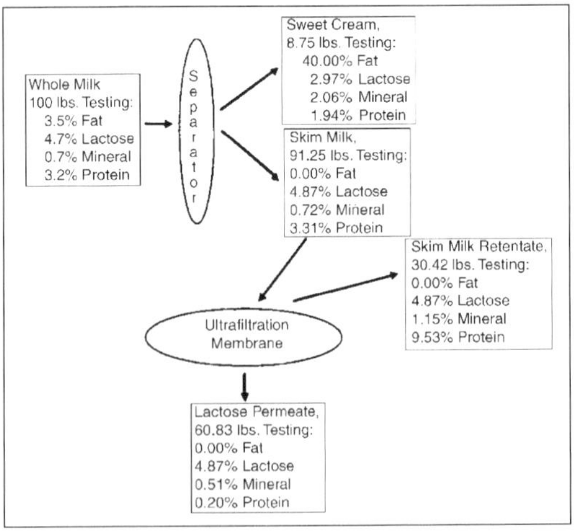

Fig. Flow and Mass Balance for the Manufacture of Skim Milk Retentate, the Building Block of New Dairy Products

Finally, even though many different varieties of butter can be made, only one basic style is popular. In contrast, many different styles of hard cheese exist because protein is capable of conveying the tastes associated with different starter cultures and manufacturing methods. Can substituting protein for lactose reduce lactose levels enough to be beneficial to lactose intolerant consumers? The enzyme lactase is responsible for the digestion of lactose in the small intestine. Individuals whose bodies produce insufficient lactase are said to be lactose intolerant. The severity of such lactose intolerance can vary from one individual to another. For individuals with only mild intolerance, the reductions achieved by a proteinfor-lactose substitution could be beneficial. This would be particularly true for yogurt, which contains other beneficial bacteria to aid digestion (U.S. National Institutes of Health). To make each of these products, skim milk retentate (SMR) is needed.

SMR is produced by ultrafiltering skim milk to create a fluid isolate with a high protein-to-lactose ratio. As a result, SMR can reduce lactose content while increasing protein content. This means that protein can be substituted for lactose, improving the nutritional profile of dairy foods as well as their taste and texture. Further, it means that protein can substituted for the texture, functionality, and mouthfeel of butterfat.

COST OF CONCEPTS

U.S. dairy processing firms evaluate the purchase of new equipment very carefully due to budgetary constraints. Detailed system-cost information was provided by four equipment manufacturers and is shown in Table.

Table. Estimated Costs to Manufacture New Dairy Products

Characteristic	High-Protein 2% Butterfat Fluid Milk	High-Protein, Lower-Lactose Ice Cream Mix	High-Protein Nonfat Yogurt Mix
System Production/Day	375,000 Ibs. milk	200,000 lbs. mix	100,000 Ibs. mix
System Capital Cost[a]	$455,000.00	$1,240,000.00	$455,000.00
10-Yr. Depreciation (312 day basis)	$145.83/day	$397.44/day	$145.83/day
Operating Cost[a]	$675.00/day	$2,025.00/day	$675.00/day
Daily Capital and Operating Cost	$820.83/day	$2,422.44/day	$820.83/day
Capital and Operating Cost per Unit	$0.22/cwt.	$1.21/cwt.	$0.82/cwt.
Added Milk Cost[b]	$1.62/cwt.	$3.59/cwt.	$1.40/cwt.
Total Added Cost	$1.84/cwt. (or $0.16/gal.)	$4.80/cwt. (or $0.41/gal. Mix)	$2.22/cwt. ($0.19/gal. Mix)
Average Retail Price	$2.50/gal.	One gallon of mix will make four half- of ice cream selling for $3 00 each.	One gallon of mix will make 17 eightgallons ounce cups of yogurt selling for $0.50 each.
Total Added Cost/ Average Retail Price	6.4%	3.4%	2.2%

Capital costs pertain to the membrane system and system hardware but exclude the cost for connection to utilities such as water, steam, and electricity. Specific capital costs include assembly, balance tanks, design engineering, electrical wiring, flow meters, gauges, installation, membrane housing, membranes, pipes, pressure gauges, process control computer, pumps, temperature recorders, and valves. Capital costs were $455,000 for the fluid milk and yogurt membrane systems and $1,240,000 for the higher-capacity ice cream system. Membrane systems are relatively small and can usually be installed within an existing building; therefore, no cost for a building has been included. Capital costs were depreciated on a straight-line basis over ten years. Operating costs include those for membrane replacement, replacement of other parts, electricity, water, steam, sanitation materials, and labour. The third and final cost area pertains to the extra cost of the milk itself.

This results from inexpensive lactose being replaced by expensive protein. Note that although the ice cream system was more expensive, since it required greater capacity due to its greater substitution of protein for lactose, only one system-cost alternative was examined for each product. Debt was not included in the cost calculations. The total added cost to make high-protein fluid milk was $0. 16 per gallon. Using an average retail price of $2.50 per gallon, the resulting cost increase relative to retail price is 6.4 per cent.

The added cost for the high-protein, lower-lactose ice cream mix was estimated at $0.41 per gallon of mix. Because of the incorporation of air, one gallon of ice cream mix will make four half-gallons of frozen ice cream. This equates to a cost increase of $0.1025 per halfgallon of frozen ice cream. Using an average retail price of $3.00 per half-gallon of ice cream, the resulting cost increase relative to retail price would be 3.4 per cent. The added cost for high-protein nonfat yogurt was estimated at $0.19 per gallon of mix. One gallon of yogurt mix will make 17 eightounce. cups of yogurt. Using an average retail price of $0.50 per cup, the resulting cost increase versus retail price would be 2.2 per cent.

SURVEY OF MILK PROCESSORS

In order to estimate the potential success of these new product concepts, a survey instrument was sent to U.S. dairy processors. Participants were informed of the particulars of the new product concept and supplied with the estimated cost of manufacturing the new dairy product. Background information was requested in a variety of areas including the respondent's opinion as to why customers purchased their existing dairy products, the importance of private-label products, the size of the firm, and the frequency of the respondent's contact with end-customers.

The survey also asked whether the firm presently employed any membrane technology for dairy purposes (only 10 per cent did so), whether the respondent thought consumers would buy the new product, and requested suggestions for increasing the probability of the new product's commercial success. A total of 179 firms were contacted, of which 63 completed the survey for a total response rate of 35 per cent.

DIFFERENCE BETWEEN DAIRY TECHNOLOGY AND DAIRY MANAGEMENT

Both the Dairy technology and Dairy management deals with same agro based field. Even though they deal within the same field, the factors and work nature differs a lot. Dairy technology includes taking care of milk giving cattle in addition to processing and gaining of milk. Dairy management includes the various principles involved in the management of dairy industries.

DAIRY TECHNOLOGY

Amongst all other industries in the country which plays a most important role in the agro based market is the Dairy industry. Dairy technology involves acquiring and processing of milk into a wide variety of dairy products. In India, more than 60% of the total population is involved in the agro related production either directly or indirectly. Professionals working in this field are usually called Dairy technologist. They are directly involved in the scientific and quality aspect of the dairy industry. These professionals work in this field in order to improve and apply diverse methods in the processing, conservation and consumption of milk and it's by products. There are number of courses available in this field in India, both at undergraduate and postgraduate levels, and for getting admission in these courses students have to undergo an entrance examination. Most of the universities organize their own admission tests.

DAIRY MANAGEMENT

With the advancements in the higher education, the professional skills are taught at different levels of study. One such diversified field is the dairy management which involves many tasks and principles. Nowadays the dairy sector has become a full-fledged field with its modern technology. The various applications of principles in the management of dairy industry are one of the main aspects of the dairy management. India is said to be the second largest milk producer in the world. In the past few years, Indian government has given more care in the dairy production and that resulted in the management of dairy farming in the professional line. The whole process taking place in this field such as processing of milk, its transportation to diverse markets, distribution, etc. all needs some professional management care. Dairy management courses are offered in many universities in the country. The course aims in providing students about the basic knowledge in the production, designing and management of the dairy industry. It also covers different areas like entrepreneurship, improvement in the milk conservation, dairy processing and management of dairy farms.

Key differentiators between Dairy Technology and Dairy Management:

- Dairy technology includes obtaining and processing of milk with various technologies to make it into different products whereas Dairy management involves the different principles and tasks in the management of dairy industry.
- Dairy technologist are only involved in the various methods and techniques needed for making diverse dairy products whereas the dairy management professionals have to manage the whole process taking place in the dairy farms, from obtaining the milk to its distribution as various by-products.

The dairy technology and dairy management are involved in the dairy field. Although they are in same field, the work nature and prospects differs in many aspects.

6

Food Packaging Technology and Dairy Product

Food and beverage packaging comprises 55% to 65% of the $130 billion value of packaging in the United States. Food processing and packaging industries spend an estimated 15% of the total variable costs on packaging materials. Industrial processing of food, reduced consumption of animal protein, importation of raw materials and ingredients to be converted in the United States, and scarcity of time to select/prepare food from fresh ingredients have enhanced innovation in food and beverage packaging.

The continued quest for innovation in food and beverage packaging is mostly driven by consumer needs and demands influenced by changing global trends, such as increased life expectancy, fewer organizations investing in food production and distribution, and regionally abundant and diverse food supply. The use of food packaging is a socioeconomic indicator of increased spending ability of the population or the gross domestic product as well as regional food availability.

This Scientific Status Summary provides an overview of the latest innovations in food packaging. It begins with a brief history of food and beverage packaging, covering the more prominent packaging developments from the past, and proceeds to more modern advances in the packaging industry.

The article then delves into current and emerging innovations in active and intelligent packaging (such as oxygen scavengers and moisture control agents), packaging mechanisms that control volatile flavours and aromas (such as flavour and odour absorbers), and cutting-edge advances in food packaging distribution (such as radio frequency identification and electronic product codes). Finally, the nano-sized components that have the potential to transform the food packaging industry.

INTELLIGENT FOOD PACKAGING

Traditional food packages are passive barriers designed to delay the adverse effects of the environment on the food product. Active packaging, however, allows packages to interact with food and the environment and play a dynamic role in food preservation. Developments in active packaging have led to advances in many areas, including delayed oxidation and controlled respiration rate, microbial growth, and moisture migration. Other active packaging technologies include carbon dioxide absorbers/emitters, odour absorbers, ethylene removers, and aroma emitters. While purge and moisture control and oxygen removal have been prominent in active packaging, purge control is the most successful commercially.

An example is the drip-absorbing pad used in the poultry industry. In addition, active packaging technology can manipulate permselectivity, which is the selective permeation of package materials to various gases. Through coating, microperforation, lamination, coextrusion, or polymer blending, permselectivity can be manipulated to modify the atmospheric concentration of gaseous compounds inside a package, relative to the oxidation or respiration kinetics of foods.

Certain nanocomposite materials can also serve as active packaging by actively preventing oxygen, carbon dioxide, and moisture from reaching food. Intelligent or smart packaging is designed to monitor and communicate information about food quality. Examples include time-temperature indicators (TTIs), ripeness indicators, biosensors, and radio frequency identification. These smart devices may be incorporated in package materials or attached to the inside or outside of a package.

As of summer 2008, the commercial application of these technologies to food packaging has been small. However, the U.S. Food and Drug Administration (FDA) recognizes TTIs in the 3rd edition of the *Fish and Fisheries Products Hazards and Control Guidance*, so their importance may increase in the seafood industry. Moreover, Wal-Mart, Home Depot and other retail outlets use radio frequency identification, so it is likely to become very prominent as a mechanism for tracking and tracing produce and others perishable commodities.

OXYGEN SCAVENGERS

The presence of oxygen in a package can trigger or accelerate oxidative reactions that result in food deterioration: Oxygen facilitates the growth of aerobic microbes and molds. Oxidative reactions result in adverse qualities such as off-odours, off-flavours, undesirable colour changes, and reduced nutritional quality. Oxygen scavengers remove oxygen (residual and/or entering), thereby retarding oxidative reactions, and they come in various forms: sachets in headspace, labels, or direct incorporation into package material and/or closures.

Oxygen scavenging compounds are mostly agents that react with oxygen to reduce its concentration. Ferrous oxide is the most commonly used scavenger. Others include ascorbic acid, sulfites, catechol, some nylons, photosensitive dyes, unsaturated hydrocarbons, ligands, and enzymes such as glucose oxidase. To prevent scavengers from acting prematurely, specialized mechanisms can trigger the scavenging reaction. For example, photosensitive dyes irradiated with ultraviolet light activate oxygen removal. Oxygen scavenging technologies have been successfully used in the meat industry.

CARBON DIOXIDE ABSORBERS AND EMITTERS

Carbon dioxide may be added for beneficial effects, for example, to suppress microbial growth in certain products such as fresh meat, poultry, cheese, and baked goods. Carbon dioxide is also used to reduce the respiration rate of fresh produce and to overcome package collapse or partial vacuum caused by oxygen scavengers. Carbon dioxide is available in various forms, such as moisture-activated bicarbonate chemicals in sachets and absorbent pads. Conversely, high levels of carbon dioxide resulting from food deterioration or oxidative reactions could cause adverse quality effects in food products. Excess carbon dioxide can be removed by using highly permeable plastics whose permeability increases with higher temperatures.

MOISTURE CONTROL AGENTS

For moisture-sensitive foods, excess moisture in packages can have detrimental results: for example, caking in powdered products, softening of crispy products such as crackers, and moistening of hygroscopic products such as sweets and candy. Conversely, too much moisture loss from food may result in product desiccation.

Moisture control agents help control water activity, thus reducing microbial growth; remove melting water from frozen products and blood or fluids from meat products; prevent condensation from fresh produce; and keep the rate of lipid oxidation in check. Desiccants such as silica gels, natural clays and calcium oxide are used with dry foods while internal humidity controllers are used for high moisture foods (for example, meat, poultry, fruits, and vegetables).

Desiccants usually take the form of internal porous sachets or perforated water-vapour barrier plastic cartridges containing desiccants. They can also be incorporated in packaging material. Humidity controllers help maintain optimum in-package relative humidity (about 85% for cut fruits and vegetables), reduce moisture loss, and retard excess moisture in headspace and interstices where microorganisms can grow. Purge absorbers remove liquid squeezed or leaking from fresh products and can be enhanced by other active additives such as oxygen scavengers, antimicrobials, pH reducers, and carbon dioxide generators.

ANTIMICROBIALS

Antimicrobials in food packaging are used to enhance quality and safety by reducing surface contamination of processed food; they are not a substitute for good sanitation practices. Antimicrobials reduce the growth rate and maximum population of microorganisms (spoilage and pathogenic) by extending the lag phase of microbes or inactivating them. Antimicrobial agents may be incorporated directly into packaging materials for slow release to the food surface or may be used in vapour form.

Research is underway on the antimicrobial properties of the following agents:

- *Silver Ions*: Silver salts function on direct contact, but they migrate slowly and react preferentially with organics. Research on the use of silver nanoparticles as antimicrobials in food packaging is ongoing, but at least 1 product has already emerged: FresherLongerTM storage containers allegedly contain silver nanoparticles infused into polypropylene base material for inhibition of growth of microorganisms.
- *Ethyl Alcohol*: Ethyl alcohol adsorbed on silica or zeolite is emitted by evaporation and is somewhat effective but leaves a secondary odour.
- *Chlorine Dioxide*: Chlorine dioxide is a gas that permeates through the packaged product. It is broadly effective against microorganisms but has adverse secondary effects such as darkening meat colour and bleaching green vegetables.
- *Nisin*: Nisin has been found to be most effective against lactic acid and Gram-positive bacteria. It acts by incorporating itself in the cytoplasmic membrane of target cells and works best in acidic conditions.
- *Organic Acids*: Organic acids such as acetic, benzoic, lactic, tartaric, and propionic are used as preservative agents.
- *Allyl Isothiocyanate*: Allyl isothiocyanate, an active component in wasabi, mustard, and horseradish, is an effective broad spectrum antimicrobial and antimycotic. However, it has strong adverse secondary odour effects in food.
- *Spice-based Essential Oils*: Spice-based essential oils have been studied for antimicrobial effects: for example, oregano oil in meat, mustard oil in bread, oregano, basil, clove, carvacol, thymol, and cinnamon.
- *Metal Oxides*: Nanoscale levels of metal oxides such as magnesium oxide and zinc oxide are being explored as antimicrobial materials for use in food packaging.

ETHYLENE ABSORBERS AND ADSORBERS

Ethylene is a natural plant hormone produced by ripening produce. It accelerates produce respiration, resulting in maturity and senescence. Removing ethylene from a package environment helps extend the shelf life of fresh produce. The most common agent of ethylene removal is potassium permanganate, which oxidizes ethylene to acetate and ethanol. Ethylene may also be removed by physical adsorption on active surfaces such as activated carbon or zeolite. Potassium permanganate is mostly supplied in sachets while other adsorbent or absorbent chemicals may be distributed as sachets or incorporated in the packaging materials.

TEMPERATURE CONTROL: SELF-HEATING AND COOLING

Self-heating packaging employs calcium or magnesium oxide and water to generate an exothermic reaction. It has been used for plastic coffee cans, military rations, and on-the-go meal platters. The heating device occupies a significant amount of volume (almost half) within the package. Self-cooling packaging involves the evaporation of an external compound that removes heat from contents (usually water that is evaporated and adsorbed on surfaces).

THE EXPANDED ROLES OF FOOD AND BEVERAGE PACKAGING

The principal function of packaging is protection and preservation from external contamination. This function involves retardation of deterioration, extension of shelf life, and maintenance of quality and safety of packaged food. Packaging protects food from environmental influences such as heat, light, the presence or absence of moisture, oxygen, pressure, enzymes, spurious odours, microorganisms, insects, dirt and dust particles, gaseous emissions, and so on. All of these cause deterioration of foods and beverages.

Prolonging shelf life involves retardation of enzymatic, microbial, and biochemical reactions through various strategies such as temperature control; moisture control; addition of chemicals such as salt, sugar, carbon dioxide, or natural acids; removal of oxygen; or a combination of these with effective packaging. Precise integration of the product, process, package, and distribution is critical to avoid recontamination. The ideal packaging material should be inert and resistant to hazards and should not allow molecular transfer from or to packaging materials. Other major functions of packaging include containment, convenience, marketing, and communication. Containment involves ensuring that a product is not intentionally spilled or dispersed. The communication function serves as the link between consumer and food processor. It contains mandatory information such as weight, source, ingredients, and now, nutritional value and cautions for use required by law. Product promotion or marketing by companies is achieved through the packages at the point of purchase.

Secondary functions of increasing importance include traceability, tamper indication, and portion control. New tracking systems enable tracking of packages though the food supply chain from source to disposal. Packages are imprinted with a universal product code to facilitate checkout and distribution control. More recent innovations used include surface variations sensed by finger tips and palms, sound/music or verbal messages, and aromas emitted as part of an active packaging spectrum. Gloss, matte, holograms, diffraction patterns, and flashing lights are also used.

RFID SYSTEMS FOR PACKAGED FOODS

Radio frequency identification (RFID) is a system that uses radio waves to track items wirelessly. RFID makes use of tags or transponders (data carriers), readers (receivers), and computer systems (software, hardware, networking, and database). The tags consist of an integrated circuit, a tag antenna, and a battery if the tag is passive (most active tags do not require battery power).

The integrated circuit contains a non-volatile memory microchip for data storage, an AC/DC converter, encode/decode modulators, a logic control, and antenna connectors. The wireless data transfer between a transponder/tag and a reader makes RFID technology far more flexible than other contact identifications, such as the barcode system and thus makes it ideal for food packaging.

The working principles of an RFID system are as follows:

- Data stored in tags are activated by readers when the objects with embedded tags enter the electromagnetic zone of a reader;
- Data are transmitted to a reader for decoding; and
- Decoded data are transferred to a computer system for further processing.

Tag frequency is related to the working principles of an RFID system (for example, magnetic coupling or electric coupling) and the reading range. Frequency depends on the type of tag, reader, and cost. The typical RFID frequencies are low frequency, high frequency, ultra high frequency, and microwave frequency.

Generally, low frequency systems have short reading ranges, slow read speeds, and lower cost while higher frequency RFID systems are utilized when longer read ranges and fast reading speeds are required. Microwave frequency requires active RFID tags.

ELECTRONIC PRODUCT CODES

In 2004, Electronic Product Code (EPC) Global Network began developing a second generation RFID protocol: EPC Class 1 version 2. The main goal of Gen 2 is to create a single global standard that is compatible with ISO standards. The tags would work in various countries that have different commercial band frequency. EPC is the most significant function of RFID contributing to commercial industry.

EPC improves the traceability of items and facilitates efficient product recall and authenticity. EPC is similar to Universal Product Code (UPC), which is commonly used in bar codes. Compared to UPC, which uses 12 digits of numbers, EPC has 64 to 256 bits of alphanumeric data. The most common EPC has 96 bits. The first obstacle for wide utilization of RFID is the cost for tags. Tags are still too expensive for use on individual primary packages.

The infrastructure required for RFID systems (including readers, database servers with communication systems, and other information technology to process huge amounts of data) is costly and needs to be shared with all users in supply chains. The global use of EPC also requires compatibility among various regulations and standards of radio frequency.

The biggest hurdle to wide utilization of EPC for RFID is the potential problems of privacy protection. A hidden reading system could collect all data from tags of items and also RFID card holders for the purpose of data stealing or data removing. Guidelines for the ethical use of RFID systems for data collecting, data handling, and system security need to be established.

RFID FOR THE FOOD INDUSTRY

RFID has recently found its way into numerous applications in the food industry, ranging from food monitoring and traceability to enhancing food safety, to improving supply chain efficiency. The major benefits of RFID technology in the food industry are greater speed and efficiency in stock rotation and better tracking of products throughout the chain, resulting in improved on-shelf availability at the retail level and enhanced forecasting. The technology is well suited for many operations in food manufacturing and supply chain management.

An RFID-based resource management system can help users handle warehouse operating orders by retrieving and analysing warehouse data, which could save time and cost. The use of RFID in the food industry is currently focused on tracking and identification. When RFID technology becomes more established in the food industry, the integration of food science knowledge will be necessary to develop the intelligent food packaging application for food quality and safety.

Some food companies have already integrated RFID into manufacturing and distribution. Retail chains such as Wal Mart and Home Depot have been testing the technology for distribution. In 2003, Wal-Mart issued a mandate requiring its top 100 suppliers to use RFID tags on all cases and pallets entering its distribution centers by 2005.

RFID compliance is a long-term project for Wal-Mart; more of its suppliers are expected to be compliant by the end of 2008. Other major players advocating RFID technology are the U.S. Department of Defence and major retailers such as Albertsons, Target, Tesco, and Marks and Spencer. RFID technology also provides security and safety benefits for food companies through tracking the origin of supplies. For example, a small California winery uses RFID to track its barrels and to enhance wine making by streamlining data collection.

The company planted RFID tags on tanks and harvesting bins, allowing better control of wine production and tracking. In addition, by attaching an RFID tag to a package, the package becomes intelligent because the stored data provide valuable information that can be stored and read by appliances. This intelligent packaging technology is also being extended to refrigeration and freezing. Appliances can communicate with the packages and identify information related to the storage of the packaged products. Despite these benefits, other factors such as cost of the technology and recycling ability need to be considered.

SUPPLY CHAIN MANAGEMENT, TRACEABILITY, AND RECALL

As a tool for tracking, RFID is a promising technology to food supply chain management. The FDA, 1307 recalls of processed foods occurred between 1999 and 2003; these recalls could have been avoided with a technology such as RFID. Exposure to risk at any one of the stages of the processed food supply chain would result in a domino-effect breakdown that could affect the smooth running of an entire supply chain.

All stages of the supply chain (from farming, processing, transportation, manufacturing, retailing, and warehousing to consumption) are equally critical to food recall problems. If RFID technology were combined with Hazard Analysis and Critical Control Point systems, the supply chain stages would be integrated, traceable, and effectively managed by food processors to reduce the number of recalls significantly. The outstanding tracing abilities of RFID tags to individual food product could enable manufacturers to audit every single phase of a product in a retail unit, monitoring correct handling, transportation, storage, and delivery.

NEW TAG FOR USE ON METAL AND METAL PACKAGING AND HIGH WATER CONTENT PRODUCTS

The RFID system is not infallible; it has some weaknesses, such as the shielding effect of metal, which affects signal transduction. Data cannot be read correctly when tags are attached to metal on the surface or inside the package. Until recently, high frequency RFID tags were used on metal beer barrels. However, low frequency RFID (125 kHz to 135 kHz) has less loss of radio signal by the metal materials in the high magnetic field of a reader compared to higher frequency signals.

Another issue is that water molecules can absorb microwave signals, resulting in signal loss or interference during data acquisition from microwave RFID tags. Since most foods contain high moisture, this signal interference requires further study to enable application of the technology in the food industry. It is worth noting that low, high, and ultra-high frequency (UHF) tags can be used in high water systems since water does not interfere with their signal; on the contrary, ice absorbs UHF radio signals.

For example, a product such as ice cream (which is technically defined as partly frozen foam with ice crystals and air bubbles occupying a majority of the space) contains about 72% frozen water and thus interferes with UHF radio signals. Ice cream manufacturers have overcome this product interference by placing tags over an air gap in the containers. Companies with diverse product lines and high levels of automation will likely encounter significant technical barriers to RFID implementation.

THE INTERSECTION OF FOOD AND GENES

Understanding of human dietary requirements results from developments in many scientific disciplines, including food science, nutrition, chemistry, biochemistry, physiology, and genetics. New research in proteomics, nutrigenomics, metabolomics, and other disciplines may help identify the biological basis by which food components promote health and wellness. Continuing and accelerating this research will reveal the effects of nutrients on the molecular-level processes in the body and document the variable effects of nutrients under different conditions.

NEW DISCIPLINES

Nutrigenomics, proteomics and metabolomics are three new disciplines that will contribute to the rapid development of functional foods. Bioinformatics is a new tool that uses computer database technology to integrate data from multiple, and sometimes disparate, disciplines. Already these disciplines and tools have improved our understanding of food science and human nutrition. Discoveries in genetics make it possible to understand the effects of nutrients in processes at the molecular level in the body and also the variable effects of dietary components on each individual. The scientific and technological discipline named nutrigenomics relies heavily on well established science and technology from the fields of genomics, proteomics, metabolomics, food science, and nutrition.

Table. Terminology and Disciplines Pertinent to Applications of Genetic Research to Nutrition and Health

Term	Definition and Function
Gene	A gene is a DNA (deoxyribonucleic acid) segment that contributes to phenotype/function as defined by HUGO. (Human Genome Organization).
Genome	Life is specified by genomes. Every organism, including humans, has a genome that contains all the biological information needed to build and maintain a living example of that organism. The biological information contained in a genome is encoded in its DNA and divided into discrete units called genes. Genes code for proteins that attach to the genome at the appropriate positions and switch on a series of reactions called gene expression.
Genomics	The characterization and study of whole genomes with respect to the DNA sequence, and the arrangement and function of genes. Further specified as: structural genomics (mapping and sequencing genes) and functional genomics (understanding the

	functions of genes, the proteins made as a result of gene activation [expression], and the interactions of those proteins).
Genotype	The genetic constitution of an organism, as distinguished from its physical appearance (its phenotype). The genetic identity of an individual that does not show as outward characteristics.
Phenotype	The physical characteristics or observable traits of an organism, *e.g.*, hair colour, weight, or the presence or absence of a disease. Phenotypic traits are not necessarily genetic.
Single Nucleotide Polymorphism (SNP) Pronounced "Snip"	Heritable, individual variations that occur in one nucleotide such that DNA and gene sequences, and ultimately proteins produced by those genes, vary from one person to the next. Differences in proteins are minor, usually on the order of one amino acid; however, effects on protein function may be significant and cause or contribute to individual differences in response to environment, such as diet and drugs. A small genetic change, or variation, that can occur within a person's DNA sequence. The genetic code is specified by the four nucleotide "letters": A (adenine), C (cytosine), T (thymine), and G (guanine). SNP variation occurs when a single nucleotide, such as an A, replaces one of the other three nucleotide letters, C, G, or T.
Polymorphism	An alternate form of a gene present in >1% of the population.
Proteomics	The study of the full set of proteins encoded and expressed by a genome, from healthy and diseased tissues. Further specified as: structural proteomics (identifying proteins by analysing amino acid sequences); molecular proteomics (studying the interactions of proteins with other proteins and cellular components); and chemical proteomics (studying the interaction of proteins with chemicals, such as drugs, nutrients and toxins).
Metabonomics or Metabolomics	Metabolite profiling measures the real outcome of the potential changes suggested by genomics and proteomics. It describes the integrated biochemical status, dynamics, interactions, and regulation of whole systems or organisms at a molecular level. Systems biology approaches present a different and broader perspective from the discrete, relatively static measurements of the past. As such, they offer new understanding of disease processes and targets and the beneficial and adverse effects of drugs; but they also bring new challenges. Exploitation of patterns rather than single indicators and the dynamic nature of metabonomics end-points suggest a doseresponse continuum and perhaps challenge both industry and regulators with the obsolescence of the crude noeffect dose/effect dose concept. Characterization of individual amenability to therapy and susceptibility to toxicity ("pharmacometabonomics") has economic and ethical implications. These opportunities and challenges will be explored in the context of the present and future roles of metabonomics in drug development.
Bioinformatics	The field of science in which biology, computer science, and information technology merge to form a single discipline based on creation and mining of extensive computerized databases of

	nucleic acid sequences, gene structures, proteins and their function, as well as environmental constituents capable of modifying gene expression. The ultimate goal of the field is to enable the discovery of new biological insights as well as to create a global perspective from which unifying principles in biology can be discerned.
Nutrigenomics	The interaction of dietary components that are nutritive (vitamins, minerals, fatty acids), bioactive (phytochemicals), or metabolites of food components (retinoic acid, eicosanoids) with genes to result in gene expression.

Briefly, nutrigenomics describes how dietary components affect the protein profile of an individual. Proteomics describes how that altered protein profile affects the biological systems of the individual, and metabolomics describes the cellular response to the changes.

The metabolite and gene expression patterns discovered with emerging bioinformatics tools may be used to monitor sequential metabolic changes in response to dietary components in functional foods, facilitating evaluation of the safety and efficacy of these components.

Nutrigenomics

For the purposes of this discussion, nutrigenomics is defined as the interaction of dietary components with genes. The dietary components of interest can be essential nutrients, other bioactive substances or metabolites of food components. On the one hand, nutrigenomics represents a logical extension of biotechnology, molecular medicine, and pharmacogenomics; on the other hand, it represents a revolution in how nutrition and diets are viewed in relation to health. Sauberlich *et al.* were among the early, dedicated pioneers who established analytical methods to assess the nutritional status of humans, using biological fluids and red and white blood cells. Additional laboratory tests for the assessment of nutritional status are needed, such as the ability to measure osteocalcin instead of relying on measurements of plasma Ca to determine calcium status. Ideally, functional assessment of nutritional status would use non-invasive biofluids and emerging highly sensitive, analytical technologies.

Proteomics

Proteomics is the study of the full set of proteins encoded and expressed by a genome. Proteomics identifies the large number of proteins in the organism, maps their interactions and analyses the proteins' biologic activities. Zhu provide a comprehensive review of available analytical techniques and their use in proteomics.

Metabolomics

Metabolomics is metabolite profiling, measuring the real outcome of the potential changes suggested by genomics and proteomics. Metabolomics investigates regulation and metabolic fluxes in individual cells or cell types. Metabonomics combines the power of high-resolution nuclear magnetic resonance with statistical data analysis of *in vivo* metabolite patterns. This technique enables rapid screening for xenobiotic toxicity, disease state, drug efficiency, nutritional status and even gene function in the "whole" organism. . This emerging investigative approach is being used to assess the adequacy and safety of xenobiotics, pharmaceutical agents, nutrients and functional phytochemicals.

FUTURE DEVELOPMENTS

Diet represents one of the key environmental factors to which our genes are exposed, from conception throughout life. Gene expression results in production of proteins that function in myriad ways within the human body, serving as enzymes, oxygen transporters, hormones, and building blocks for cells throughout the body. Simply put, gene expression governs our existence.

Nutrients, in turn, govern the concentration of different proteins in different organs by functioning as regulators of gene transcription and translation, nuclear RNA (ribonucleic acid) processing, messenger RNA (mRNA) stability, and mRNA degradation. The intensity of a dietary signal and the subsequent response can vary with the amount of a food component consumed and the frequency with which it is ingested. The developmental age of the individual also may determine which genes are influenced.

Although an exhaustive review of the scientific literature is beyond the scope of this report, Tables provide an overview. As summarized in Table, research has shown that nutrients affect gene expression and formation of various proteins at discrete points in the processes leading to enzymes, structural proteins, and other chemicals on which life depends.

Thus, the amount—and even the form—of nutrients present during gene-expression can affect the synthesis of protein, resulting in less of a protein being produced, production of a less than optimally functional form, or no protein at all. Each of those possibilities exists due to the hereditary form of genes present and whether the genes are normal or contain polymorphisms that affect gene expression.

Table. Gene Expression Processes Leading to Protein Formation and Selected Nutrient Regulators in the Process

Gene Expression Sequence	Nutrient Regulator
Gene transcription	Fatty acids, glucose, cholesterol, amino acids, zinc, bioactive components
mRNA processing	Methionine, choline,

	vitamins B-6 and B-12, fatty acids
mRNA stability	Amino acids, vitamin D, calcium
mRNA translation	Glucose, fatty acids, minerals, amino acids, choline, conjugated linoleic acid (CLA)
Post-translational modification	Minerals and vitamin cofactors
Protein transport to functional location	Vitamins, minerals

Studies designed to identify specific effects of diet on phenotypic expression of biochemical components that determine health have resulted in tantalizing suggestions for dietary interventions designed to modify gene expression.

Table. Examples of Nutrient Involvement in Gene Expression and Potential Phenotypic Results

Nutrient Deficiency	Phenotypic Expression
Folate	Elevated homocysteine (cardiovascular disease), neural tube defects, central nervous system dysfunction
Fatty acids	Cognitive function (depression), obesity, Inflammation
Vitamin D	Osteoporosis

Nutrients serve as substrates, cofactors, or coenzymes for metabolic processes that are familiar from traditional nutritional research and epidemiological observation. New genetic research techniques are finding that nutrients also regulate the genes whose expression leads to enzymes, transporters, and structural elements that comprise the living, functioning organism. The premise that foods consumed during the first weeks and months of life may have permanent effects on metabolism is not new.

In fact, the relationship was first recognized more than 40 years ago. Further studies in humans and animals showed permanent effects of early diet on adult metabolism, cognitive function, and body composition through activation or suppression of gene expression, or turning genes "on" or "off". Ample scientific evidence demonstrates that diet is a significant environmental determinant, if not the key determinant, of population or individual genetic expression. Those effects can be overt, such as the effects seen in vitamin deficiency diseases, or more subtle and complex, as in the manifestation of type 2 diabetes, predisposition to obesity, and other chronic diseases. For example, epidemiological surveys of adults born after prenatal exposure to famine and biochemical investigations of insulin resistance in low-birth-weight children both show a genetic basis for the observed association between low birth weight and an increased risk of developing type 2 diabetes later in life.

The predisposing genetic changes have been shown to occur *in utero*. Although scientists knew such a relationship existed between early diet and gene expression, they were unable to understand how the effect took place. Now, the integration of genomics and nutrition is providing an emerging understanding—at the molecular level—of how diet affects gene expression. This new understanding opens the door for many potential nutritional interventions, both in food composition and in food selection.

The health consequences of the interaction between an individual's diet and his or her genetic makeup have been repeatedly demonstrated. In fact, some life-threatening errors of metabolism have been successfully managed with diet modification. For example, galactosemia, a genetic disorder characterized by an inability to convert galactose to glucose, is usually discovered in infants fed milk shortly after birth because milk contains a large quantity of galactose. If not treated, galactosemia can result in cataracts, enlarged liver and spleen, and mental retardation.

It is treated by lifelong elimination of milk and other dairy products from the diet. Another example of an inborn error of metabolism, phenylketonuria (PKU) is caused by an enzyme defect in the liver that breaks down phenylalanine. As a result, phenylalanine builds up in the body, causing mental retardation. Although PKU cannot be prevented, if detected early in life, it can be successfully treated by consuming a diet low in phenylalanine.

The Human Genome Project and associated programmes have provided the groundwork for scientists to be able to pursue key questions, such as: What DNA variants underlie disease and health? How does environment interact with genes, subjecting some individuals to intractable obesity, cardiovascular disease or Alzheimer's disease at early ages, while others have a long life with little or no disease? Genetic factors may confer susceptibility or resistance to a disease and may determine the severity or progression of disease.

Since we do not yet know all of the factors involved in these intricate pathways, researchers have found it difficult to develop screening tests for most diseases and disorders. Today this can be solved by studying stretches of DNA that have been found to harbor a single nucleotide polymorphism (SNP) associated with a disease trait, researchers may begin to find relevant genes associated with a disease and variable response to dietary components. It is already possible to identify individuals with an SNP profile that predicts variable cardiovascular health status in response to diets with a particular fat composition.

Defining and understanding the role of genetic factors in disease also will allow researchers to better evaluate the role that non-genetic factors—such as behaviour, diet, lifestyle and physical activity—have on disease. While the SNPs or polymorphisms that appear to be associated with some diseases can be identified, a substantial amount of biological research remains to be completed to unequivocally link, in a cause-effect equation, the phenotypic expression of health or disease in response to intake of a specific nutrient or bioactive component.

Experimental results show that individuals whose genetic makeup contains particular SNPs may respond to dietary components in ways that result in gene expression that leads to disease phenotypes. The challenges facing nutrigenomics are similar to those encountered in drug development. Many common diseases are not caused by a genetic variation within a single gene. Instead, diseases are caused by complex interactions among multiple genes, in conjunction with environmental and lifestyle factors.

Although both environmental and lifestyle factors contribute tremendously to disease risk, their relative contributions and effects are currently difficult to measure and evaluate. Now that the human genome has been catalogued, the race is on to determine the functional significance of each gene, understand the complex functional networks and control mechanisms, and figure out the role that genotype and environment play in determining the phenotype of an individual.

Functional studies to date have largely evaluated one gene at a time. However, to truly understand the biology of processes directed by genes, researchers need to simultaneously study functional interactions, networks, and pathways. With enough data and proper bioinformatics tools, scientists will be able to model the genetic circuitry to identify interventions that can optimize biological outcomes through health and wellness lifestyle choices such as diet.

UNLOCKING THE SECRETS OF FUNCTIONAL FOOD COMPONENTS

Food technology and improved nutrition have played critical roles in the dramatic increase in life expectancy over the past 200 years, but the impact of diet on health is much broader than basic nutrition. A growing body of evidence documents positive health benefits from food components not considered nutrients in the traditional definition.

Scientific advances have allowed researchers to better characterize the biological basis of disease states, understand the metabolism of food at the cellular level, and identify the role of bioactive components in food and assess their impact on metabolic processes. New powerful analytical tools can enable scientists to unlock the biological functions of vast numbers of food components and their role in disease prevention and health promotion.

Functional foods can take many forms. Some may be conventional foods with bioactive components that can now be identified and linked to positive health outcomes. Some may be fortified or enhanced foods, specifically created to reduce disease risk for a certain group of people. Consumers can already select from a wide spectrum of foods that contain functional components either inherently or via fortification.

Health benefits may result from increasing the consumption of substances already part of an individual's diet or from adding new substances to an individual's diet. As additional bioactive components are identified, the opportunities for developing functional foods will be broad. Foods that naturally provide a bioactive substance may be enhanced to increase the level present in the food.

Alternately, foods that do not naturally contain a substance can be fortified to provide consumers with a broader selection of food sources for a particular component and its health benefit. Areas for research include better understanding the role and optimal levels of traditional nutrients for specific segments of the population, as well as identifying bioactive substances present in foods and establishing optimal levels. Early nutrition research focused on the range of vitamin and mineral intakes necessary to prevent frank deficiencies.

Now, researchers are investigating the optimum intake levels for traditional nutrients and the differences for various subpopulations. Understanding the role of nutrients at the molecular level will result in even more specific recommended dietary allowances for different population subgroups. Similar research is needed to identify the role of other bioactive food components, an area of research that is still in its infancy.

Only recently, several government agencies have begun developing a standard definition for "bioactive" food components. Research has proven that food and isolated food components can reduce the risk of disease, from the effect of vitamin A from eggs on blindness to the effect of zinc from high-protein foods on the immune system. Some examples of foods that may be considered functional foods include calcium-fortified orange juice, phytosterol/stanol-fortified spreads and juices, folate-enriched foods, soluble oat fibre, cranberry, and soy.

Table. Examples of Functional Food Components Currently Marketed

Functional Component	Health Benefits	U.S. Regulatory Status of Claims
Soluble oat fibre	Coronary heart disease	FDA approved health claim
Soy protein	Coronary heart disease	FDA approved health claim
Phytosterol/stanol esters	Coronary heart disease	FDA approved health claim (interim final rule)
Calcium	Osteoporosis	FDA approved health claim
Folate-enriched foods	Neural tube defects	FDA approved health claim

Research currently underway at academic, industry and government facilities will reveal how a myriad of substances can be used as functional food components. Although additional research is necessary to validate efficacy and establish appropriate dietary levels, researchers have identified functional food components that may improve memory, reduce arthritis, reduce cardiovascular disease and provide other benefits typically associated with drugs. In addition, new technologies will provide opportunities to produce bioactive food components from nontraditional sources. For example, Abbadi developed transgenic plant oils enriched with very long chain polyunsaturated fatty acids.

Other research has produced stearidonic acid in canola seeds to provide another source of omega-3 fatty acids in the diet. Emerging science requires that we broaden our frame of reference to take full advantage of these new discoveries. Foods may be developed to promote the expression of specific metabolites, reducing or preventing common diseases that afflict consumers with a specific genotype. Consumers might select functional foods and tailor their diets to meet changing health goals and different requirements at different ages. Future benefits might include functional foods for increased energy, mental alertness, and better sleep.

SHIFTING THE PARADIGM FOR HEALTH AND WELLNESS

A growing number of consumers perceive the ability to control their health by improving their present health and/or hedging against aging and future disease. These consumers create a demand for food products with enhanced characteristics and associated health benefits. In one study, 93% of consumers believed certain foods have health benefits that may reduce the risk of disease or other health concerns. In addition, 85% expressed interest in learning more about the health benefits offered by functional foods. Using foods to provide benefits beyond preventing deficiency diseases is a logical extension of traditional nutritional interventions. Nonetheless, such an extension requires changes in not only the foods themselves, but also their regulation and marketing—truly a paradigm shift. Creating a scientifically valid distinction between food and medicine has never been easy. Centuries ago, Hippocrates advised, "Let food be thy medicine and medicine be thy food." Early nutrition research resulted in cures for numerous widespread deficiency-based diseases. Recent scientific advances have further blurred the line between food and medicine, as scientists identify bioactive food components that can reduce the risk of chronic disease, improve quality of life, and promote proper growth and development.

The Traditional Paradigm

Traditional fortification of foods with vitamins and minerals has been accepted by consumers and regulators, but consumers should recognize the clear distinction between the use and purpose of foods vs. drugs.

Table. Benefits and Risks of Foods vs. Drugs

Food and Food Components	Drugs
Energy/nutrition/necessary for life	Treatment of disease
Life long use and benefits	Immediate effect
All populations	Target population
Safea[a]	Benefit > risk
Consumer selects	Health provider prescribes

Notes:

Adapted from Yetley, 1996

[a]Safe when consumed as a food, but with a potential increase in risk as the component levels increase. Safety evaluation will be conducted to identify the limits

Food has traditionally been viewed as a means of providing normal growth and development. Regulatory policies were established to replace nutrients lost during processing and, in some cases, to prevent nutrient deficiencies in the population. Federal policies have generally required that other diseases be treated and managed through the use of drugs.

A New Paradigm

A new self-care paradigm recognizes that foods can provide health benefits that can co-exist with traditional medical approaches to disease treatment. Science has clearly demonstrated additional dietary roles in reducing disease risk, and consumers have learned that food has a greater impact on health than previously known. At the same time, consumers recognize problems with the current healthcare system, perceiving that it is often expensive, time-constrained, and impersonal. Functional foods fit into a continuum that ranges from health maintenance/promotion to disease treatment.

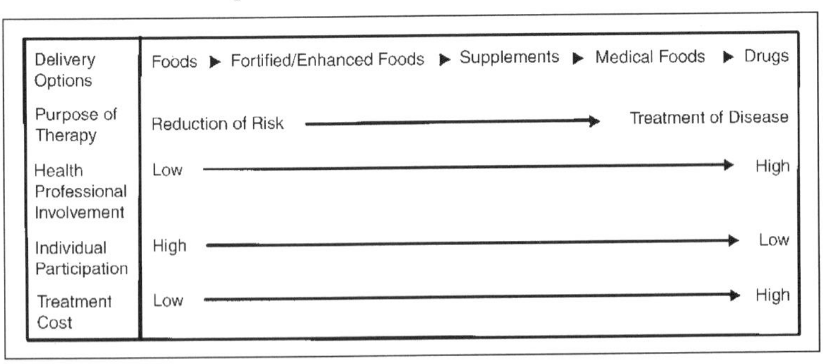

Fig. Role of Functional Foods in Health Care Continuum

On one end of the continuum are public health programmes aimed at reducing disease risk in a large segment of the population through self-directed lifestyle changes. The other end of the continuum is individualized treatment of disease by health care professionals using drugs and other medical interventions. Although the health professional involvement is low in self-directed treatment relative to individualized treatment, an important educational component remains.

New functional foods will continue to expand the continuum, providing additional options for consumers. There is a role for all aspects of this paradigm in our health care system. Functional foods should be integral components of established public health programmes to reduce the risk of specific diseases.

Treatment and prevention of coronary heart disease (CHD) provides an example of this paradigm shift. In the past, recommendations for treating hypercholesterolemia, one of the risk factors for CHD, included dietary and lifestyle interventions along with medication. The dietary and lifestyle interventions included reducing intake of saturated fat and cholesterol, quitting smoking, increasing regular physical activity, and maintaining a healthy body weight. These recommendations, often in conjunction with medication, have been effective strategies for managing heart disease.

The most recent clinical guidelines for treatment of coronary heart disease include therapeutic dietary options for reducing low density lipoproteins (LDL) by consuming specific foods, such as those that contain plant stanols/sterols, increasing intake of soluble fibre, and reducing intake of trans fatty acids. Several food components currently under study may provide additional dietary options in the prevention and treatment of CHD.

TAILORING DIETS FOR SPECIAL NEEDS

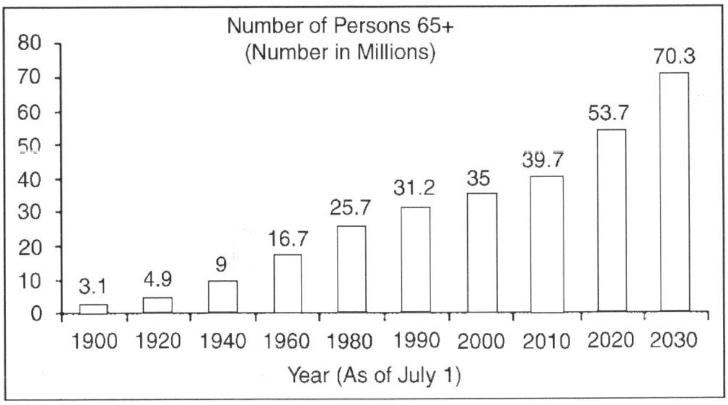

Fig. Projected Increase in Number of Elderly Individuals

Functional foods can address many consumer needs within the new paradigm when used as part of a diet tailored to address the special health needs of a specific group of consumers. In addition to those with needs because of chronic medical conditions, other groups with special needs include women of childbearing age, adolescent girls and boys, athletes, military personnel, and the elderly. For example, improving the health of the elderly in cost effective and consumer-acceptable ways will become even more urgent as the population of individuals 65 years of age and over increases by approximately 50% during the next 27 years.

The Institute of Medicine reported that poor nutritional status is a major issue for older citizens and that at least four health conditions would benefit from nutritional intervention in either "preventative or treatment modes." Some functional foods are already available for each of these purposes, but more are needed. Many elderly individuals may benefit by expanding their use of functional foods and supplements, particularly where new research can guide their selection of those foods to meet specific needs. It would be unreasonable to expect functional foods to address all of the elderly's medical needs, but functional foods can improve health and wellness, minimize costs, and provide consumers with greater control.

ENCOURAGING THE DEVELOPMENT OF FUNCTIONAL FOODS

As research provides clear evidence of relationships between dietary components and health benefits, the challenge has just begun. Scientific, regulatory, and business frameworks must be in place to evaluate the data for efficacy and safety, ensure effective regulatory oversight, communicate the findings to consumers, and provide incentives that encourage research and development of these novel food products. This report recommends modifications to the existing efficacy and safety evaluation process to ensure a sound scientific underpinning for each proposed functional food, while providing clear information to consumers.

Corresponding improvements in the regulatory oversight of new functional components also are proposed. These changes must be implemented now to protect consumer confidence in the safety of the food supply and to encourage the food industry to invest in the development of new functional foods. Science is moving rapidly; industry and government must also move rapidly to ensure that the results are translated into benefits for the consumer.

The functional foods currently available represent only a fraction of the potential opportunities for consumers to manage health through diet. Traditional definitions and arbitrary distinctions between food and medicine should not prevent consumer access to knowledge about the benefits of incorporating functional foods into their diets. Likewise, the framework for providing strong regulatory oversight should not present unnecessary barriers to the development and marketing of functional foods. Where existing terminology and regulatory frameworks are inadequate to address the full scope of benefits and opportunities for functional foods, the terminology and the frameworks must be modified.

Developing a new functional food is an expensive process. Food companies have traditionally funded research for new food product formulations but for functional foods, the stakes are higher—for both food companies and consumers.

Government investment in basic and applied research will promote the development of functional foods, but additional incentives are needed to reward private companies that pioneer new health claims.

The research required for a functional food to meet scientific standards for efficacy and safety is a substantial investment, but currently the return on that investment is not exclusive to the company that conducted the research and developed the initial regulatory petition. As soon as the health claim is adequately documented, competing companies can use the claim. Incentives, such as a period of exclusivity or tax incentives, would encourage food companies to pursue functional food development by ensuring a profitable return on successful products.

SUSTAINABLE FOOD PACKAGING

Three concepts populate the key areas of change within food packaging. The first is the trend towards more sustainable packaging. While there are multiple definitions of sustainable packaging, the Sustainable Packaging Coalition, an international consortium of more than 200 industry members, offers the most accepted definition.

Sustainable packaging is characterized by the following criteria:

- It is beneficial, safe, and healthy for individuals and communities throughout its life cycle.
- It meets market criteria for performance and cost.
- It is sourced, manufactured, transported, and recycled using renewable energy.
- It maximizes the use of renewable or recycled source materials.
- It is manufactured using clean production technologies and best practices.
- It is made from materials healthy in all probable end-of-life scenarios.
- It is designed to optimize materials and energy.
- It is recovered effectively and used in biological and/or industrial cradle-to-cradle cycles.

Sustainability initiatives led by global legislation, retailers, and corporations guide package material choices, design, and food packaging sales for food packaging professionals. The revised 1997 European Commission's Packaging and Packaging Waste Directive, the 2007 REACH (Registration, Evaluation, and Authorization of Chemicals), and the BS EN 13432 standard (which defines compost-ability, degradability, and biodegradability) are examples of effective global legislative guides in a majority of the Group of 8 countries (Canada, France, Germany, Italy, Japan, Russia, United Kingdom, United States, and the European Union), BRIC (Brazil, Russia, India, and China), and the developing world. In retail, IKEA and United Kingdom retailers have been long-term promoters of sustainable packaging and have launched impressive initiatives such as Plan A byMarks and Spencer, which defines food packaging material use. Also, Wal-Mart, the world's largest retailer, entered the sustainable packaging arena in 2006.

Introduced in 2007, the Wal-Mart scorecard ranks packages compared with category competitors, based on environmental scores.

Finally, because a packaging material's source has been shown to be a defining factor in the final package's sustainability, corporations and environmental coalitions are working together to reduce the effect of packaging materials on global sources. International paper companies, the World Wildlife Fund, the Sustainable Forestry Initiative, and the Forest Stewardship Council verify the sources of wood for use in packaging.

Groups such as the Confederation of European Paper Industries (CEPI) in the European Union are addressing solutions to the remaining high environmental impact of paper, which is heavily dependent on water, gas, and energy. Other material-based working groups are taking similar actions.

USE OF PACKAGING SUPPLIER RELATIONSHIPS FOR COMPETITIVE ADVANTAGE

The second key area of change in food packaging is the use of packaging-development value chain relationships for competitive advantage. Suppliers to the food packaging industry are adding value in relationships within the traditional package development value chain. This chain extends from raw material generation, to conversion, to production and distribution, to retail, to consumer use and disposal.

But the formerly linear nature of this chain has become an integrated sphere that enables cross-fertilization of ideas fromvarious supply-chain functions directly to the food manufacturer. Investing in packaging supply chain relationships offers opportunities for focus, innovation, and technology transfer for a competitive advantage. For example, Starbucks engaged their packaging value chain—Mississippi River Corp., MeadWestvaco, and the Solo Cup Co.—to create an FDA-approved 10% recycled-fibre coffee cup.

In addition, Green Mountain Coffee Roasters used its supply chain effectively when it introduced a compost able cup (the Ecotainer) with members of their supplier chain: DaniMer Scientific, NatureWorks, and International Paper. Also, Naturipe has used the value chain optimization to enable consistent packaging to stores from over 50 global locations.

EVOLUTION OF FOOD SERVICE PACKAGING

The third and final key area of change in food packaging is the evolution of food service packaging. Food service has grown to become a major part of consumer spending. As this trend increases, packaging plays a key role in ensuring food safety and providing convenience to consumers. For instance, proper package labeling allows food preparers to know the source of a food, its proper holding temperature, and the adequate cooking needed.

Ease of package opening allows for fewer utensils needed to open packages, which lessens contamination. While tracking and shelf-life extension technologies are employed in the food service industry to reduce the risk of food borne illness, proper heating and heat retention continue to be challenges. Consequently, packaging's role in ensuring proper heating and heat retention continues to increase. For example, CuliDish is a new product that uses varying levels of aluminum within a tray to package foods that require heating together with those that do not require heating.

The tray allows foods requiring high heat to be heated in the microwave at the same time with foods that do not require heat (such as salads). The food can be served in the same tray, thus reducing handling. More innovations in the areas of heat and heat retention are expected to assist in reducing food safety risks associated with improper cooking.

Two major convenience trends—meals eaten in transit and multi-component meals—have also advanced the food service packaging industry. The popularity of meals eaten in transit is evident since only 60% of meals are prepared and eaten at home and 20% of consumers eat on their way to another location instead of in a fixed position. On-the-go food consumption has resulted in packaging that contains a greater variety of foods.

Technologies that support this trend include edible films to enrobe food particles and edible wraps that peel off, allowing consumers to eat numerous foods while in transit. Food package design innovations such as modular folding cartons with flip-off lids, pouches that are easily opened or have a seatbelt flap to hold food close to consumers, and reusable packaging have contributed to the increase of eating while on the go. The growth of multi-component meals in food service stems from multiple types of foods being ordered at quick-service restaurants. This innovation allows for ease of food service preparation and less waste.

Multi-component packaging also provides consumers with a presentation platform that makes multi-component foods easy to consume. Examples include reusable trays that clip together into 1 larger serving tray such as that used in the Les Petits Grande line of products; KFC's triple dip strip cartons, which allow consumers to select from 1 of 3 dips when dipping chicken strips; and the folding tray that allows consumers to carry 2 cups of coffee with 1 hand.

FLAVOUR AND ODOUR ABSORBERS

In packaged foods, flavour and odour absorbers take in unwanted gaseous molecules such as volatile package ingredients, chemical metabolites of foods, microbial metabolites, respiration products, or off-flavours in raw foods. Examples are sulfurous compounds and amines produced biochemically from protein degradation, aldehydes and ketones produced from lipid oxidation or anaerobic glycolysis, and bitter taste compounds in grapefruit juice. Flavour- and odour-absorbing systems use the same mass transfer mechanism as flavour scalping to remove off-characteristics and are typically available as films, sachets, tapes/labels, and trays.

Flavour and odour absorbers are usually placed inside packages or combined with other flavour permeable materials. Although scavenging of malodourous constituents is recommended to improve the quality of packaged foods, this technology should not be used to mask off-odours produced by hazardous microorganisms that could place consumers at risk. Absorption is the dominant mass transfer phenomenon when porous materials absorb flavour molecules.

Porous or fineparticulate materials have extremely large surface areas; therefore, even with low absorption rates, the overall absorption is very significant. Porous materials—such as zeolites, clays, molecular sieves, active carbon, maltodextrin, and cyclodextrin—have been found to increase flavour. Besides the absorption rate, the maximum absorption, which is also a significant factor, can be controlled by the thickness of films, with thicker films absorbing more flavours. The film thickness does not alter absorption kinetics but relates to the total amount of absorption.

The overall absorption rate that includes the surface adsorption rate and in-structure diffusion rate is affected by molecular size of flavours, polarity of flavour molecules and packaging materials, porosity of packaging materials, storage temperature, morphology of plastic packaging materials (that is, glass transition temperature, crystallinity, and free volume), and relative humidity.

Polar plastic materials or corona treatment of film surface and/or high-barrier (or high-density and more-crystalline) materials can be used to prevent nonpolar flavour absorption. When nonporous plastic materials absorb flavour molecules without the occurrence of chemical reactions, the increase in adsorption kinetics and diffusion kinetics will accelerate the absorption rate.

However, when the flavour absorption results in chemical reactions with the plastic material, reaction kinetics as well as adsorption and diffusion kinetics significantly affect the off-flavour elimination capacity. Alkali chemicals such as ammonia and amines can be absorbed and neutralized by packaging materials containing acidic ingredients. This acid–alkali reaction mechanism between volatile chemicals and polymer ingredients has been utilized to eliminate various alkali chemicals and aldehydes.

HIGH CHEMICAL BARRIER MATERIAL INNOVATIONS

High-barrier packaging can significantly reduce adsorption, desorption, and diffusion of gases and liquids to maintain the quality of food. It also prevents the penetration of other molecules such as oxygen, pressurized liquid or gas, and water vapour, which are generally undesirable for food preservation. There are various procedures to enhance the barrier property of packaging materials or packages.

Barrier properties can be improved by combining the package materials with other high-barrier materials through polymer blending, coating, lamination, or metallization. The morphology of the blend relates to its permeability.

Laminar structure (such as coating or lamination) of high-barrier materials on packaging material decreases the permeability linearly with respect to the square thickness; also, blending with platelets or droplets of high-barrier materials reduces permeability but is less effective than coating or lamination at the same mass as that of highbarrier materials. Commercial examples are polyvinylidene chloride (PVdC) coating on oriented polypropylene (OPP), polyethylene terephthalate (PET) lamination on coextruded polypropylene/ polyethylene, and aluminum-metallization on PET.

Table. Oxygen Transmission Rate (OTR in cm^3 m^{-2} d^{-1} atm^{-1} at 23 °C, 50% RH) and Water Vapour Transmission Rate (WVTR in g m^{-2} d^{-1} at 23 °C, 75% RH) of Composite Films based on 12 μm PET Films

Film	OTR	WVTR	Specification (μm)
PET	110	15	12
PET/PE	0.93 to 1.24	0.248 to 0.372	12/50
PET/PVDC/PE	0.33	0.132	12/4/50
PET/PVAL/PE	0.13	0.26 to 0.39	12/3/50
PET/EVOH/PE	0.06	0.134 to 0.268	12/5/50
PET/Al-met/PE	0.06 to 0.12	0.006 to 0.03	12/-/50
PET/SiO$_x$	0.006 to 0.06	0.0024 to 0.06	12/-
PET/Al-foil/PE	0	0	12/9/50

Notes:

PE = polyethylene low density; PVDC = poly(vinylidene chloride); PVAL = poly(vinyl alcohol); EVOH = ethylene vinyl alcohol; Al-met = aluminum metallization; SiO$_x$ = silicon oxide; Al-foil = aluminum foil.

Table shows barrier properties of commercially available laminated (or coated) PET films. Orientation of crystalline polymers such as bi-axially oriented polypropylene (BOPP) and biaxially oriented PET (BOPET) produces lamella structure of polymers. Other examples are blends of polymers with planar clay particles, a lamella blend structure, and a mixture of beeswax in edible polymer as particulate system films.

Other innovative technologies for barrier property enhancement with commercial feasibility include transparent vacuum-deposited or plasma-deposited coating of silica oxide on PET films, epoxy spray on PET bottles, and composites of plastics with nanoparticles.

FACTORS AFFECTING FOOD PACKAGING INTERACTIONS AND BARRIER PROPERTIES

Regardless of the direction of mass transfer (for example, whether flavour absorption or release) or the intent of the mass transfer (for example, whether to achieve desirable transfer or to prevent an undesirable transfer), various factors of food, packaging, and distribution affect the mass transfer kinetics and amount. The transfer of chemicals is principally derived by the concentration difference in food and packaging materials. A small concentration gradient results in a little transfer while a large gradient results in transfer of a large amount of the compound at a fast rate.

The nature of food is also an important factor: for example, food ingredients like lipids and flavours act as solvents of plastic materials, making them soft, even barrier plastic materials. Lipid and moisture content generally control the transfer of chemicals significantly. Storage temperature, storage duration, and contact surface are also important factors. A high storage temperature accelerates transfer while longer storage duration raises the amount of transfer.

In addition, nonvolatile migrants and flavours can be transferred when the food and packaging are in contact. In such cases, migration time is related to the cumulative duration of contact of migrants to food. Moreover, a package's entire surface area may affect the migration of volatiles and flavours. The use of barrier packaging or inert materials always reduces this type of transfer. Therefore, simple lamination or coating on packaging material surfaces is a very practical way to modify the barrier properties and, consequently, to control the chemical interactions between food and the package. Use of inert packaging materials with hermetic seals can also be used to contain volatile flavours.

7

Raw Milk Transport and Storage

In developed countries, raw milk on the farm is usually cooled quickly and stored in refrigerated bulk tanks at <7 °C prior to collection. Collection by insulated tanker is often on alternate days, or sometimes less frequently, and therefore some of the milk in the tank could be 48 hours old at the time of collection. Temperature control is thus critical to minimise microbial growth, and tanker drivers are usually permitted to refuse milk stored at too high a temperature, or which has an abnormal appearance or odour.

Bacterial numbers in the milk may increase during transport, either as a result of contamination from inadequately cleaned tankers or from the growth of psychrotrophic organisms, particularly *Pseudomonas* spp.. Milk temperature and duration of the transport stage are therefore important factors. On arrival at the processing site, the milk is transferred to bulk storage tanks, or silos, prior to processing.

The milk may be stored in the silos for 2-3 days, and further growth of psychrotrophic bacteria is likely during this period. The degree of growth is dependent on the initial microbial load, and the storage time and temperature. Pseudomonads are the predominant organisms present in stored raw milk, with *Pseudomonas fluorescens*, *Pseudomonas fragi*, and *Pseudomonas lundensis* being commonly isolated, but Enterobacteriaceae, *Flavobacterium*, *Alcaligenes*, and Gram-positive species can also be found.

The growth of psychrotrophic bacteria may also be accompanied by the production of heat-stable, extracellular proteolytic and lipolytic enzymes. These enzymes are often capable of surviving pasteurisation and, in some cases, ultra high temperature processing, and they may subsequently cause spoilage in the processed milk. A number of techniques have been used to limit the growth of psychrotrophs during raw milk storage.

RAW MILK

Before the adoption of routine pasteurisation, milk was an important vehicle for the transmission of a wide range of diseases, including typhoid, brucellosis and diphtheria. Pasteurisation and improvements in veterinary medicine have seen a very large reduction in the incidence of such traditionally milkborne diseases. However, raw milk may still contain a very wide range of pathogens, including *Salmonella* spp., *E. coli* O157, *L. monocytogenes* and *Campylobacter* spp. derived from the milk animals, the environment or from farm workers and milking equipment. Pathogens may be present even in hygienically produced milk of generally good microbiological quality. In short, raw milk is a potentially hazardous product, the microbiological safety of which cannot be assured without the use of pasteurisation or an equivalent process. Recent milk-associated outbreaks of infectious intestinal disease in the UK have been shown to be caused mainly by unpasteurised or inadequately pasteurised milk products.

PASTEURISED MILK PRODUCTS

Salmonella

Salmonellae are not able to survive the typical minimum pasteurisation processes generally prescribed in legislation. Therefore, their presence indicates that the process has not been carried out effectively, or that post-process contamination has occurred. For example, an outbreak of salmonellosis in Kentucky in 1984 was associated with pasteurised milk, but an investigation of the dairy concerned showed that pasteurisation temperatures were inadequate, and could have been as low as 54.5°C for 30 minutes.

An outbreak caused by *Salmonella braenderup* in the UK in 1986 was also associated with pasteurised milk, and on this occasion the pasteuriser was found to be poorly designed and incorrectly operated, probably resulting in the application of an inadequate heat treatment. In 1985, one of the largest outbreaks of salmonellosis in US history occurred in Illinois. Almost 200,000 people were affected, and were associated with pasteurised low-fat 2% milk contaminated with *S. typhimurium*. Investigations at the dairy plant involved revealed no evidence of inadequate pasteurisation, and the outbreak strain was not found to be abnormally heat resistant.

Although the cause of the outbreak has never been completely explained, the investigation did discover a possible cross-connection between raw and pasteurised milk, which may have been the source of contamination. In 1998, an outbreak of salmonellosis in Lancashire, caused by a multiresistant strain of *S. typhimurium* DT104, affected 86 people. This outbreak was also linked to defective pasteurisation of milk at a dairy on a local farm.

Consumption of raw milk or raw milk products have been responsible for 62 and 29 cases of diarrheal illness caused by *S. typhimurium* in 2003 and 2007, respectively, in the states of Ohio and Pennsylvania. Since salmonellae are occasional contaminants of raw milk, they may sometimes enter the processing environment. It is very important that contamination of the post-pasteurisation plant is not allowed to occur and effective precautions and monitoring procedures, based on HACCP principles, are necessary to prevent this.

Campylobacter Spp.

Campylobacter spp. are not capable of surviving milk pasteurisation treatments, and cannot grow in raw or pasteurised milk, although they are able to survive for long periods in milk at refrigeration temperatures. Nonetheless, outbreaks of campylo-bacteriosis associated with pasteurised milk have occurred. For example, a large outbreak in the UK in 1979 caused by *Campylobacter jejuni* was estimated to have affected at least 2,500 schoolchildren, and was associated with free milk provided in schools. Although conclusive evidence was absent, it seems likely that raw milk may have bypassed the pasteurisation process.

A more recent outbreak in 2001 involved 75 people and was linked to the consumption of unpasteurised milk procured thorough a cow leasing programme. Birds are known to be an important reservoir of *Campylobacter* infection, and the tendency of some birds to peck through the foil tops of doorstep-delivered milk bottles is becoming recognised as an important source of infection in parts of the UK.

Some individual cases have been attributed to this cause, and, in one instance in 1990, the organism was isolated from the beaks of jackdaws and magpies as well as the contaminated milk. More recently, an outbreak thought to be associated with bird-pecked milk was reported.

Listeria Monocytogenes

There has been some discussion regarding the potential for *L. monocytogenes* in milk to survive pasteurisation. An outbreak of listeriosis in Massachusetts during 1983 resulted in 49 cases, 14 of whom subsequently died. Epidemiological evidence strongly suggested an association with consumption of pasteurised whole and low-fat milk, although this could not be confirmed microbiologically. The investigation failed to reveal any evidence of inadequate pasteurisation and the organism could not be found in environmental samples in the dairy, suggesting that postprocess contamination was unlikely.

However, samples of raw milk taken from farms supplying the plant were found to be positive for *L. monocytogenes* serotype 4b, and the investigators concluded that survival of some organisms through pasteurisation was the most likely cause of the outbreak. Three deaths and a miscarriage in Boston, USA between 2007-8 have been linked to presence of *Listeria* in pasteurised milk. So far, investigations have found nothing wrong with its pasteurisation process.

Furthermore, in a survcy of pasteurised milk conducted in Spain, *L. monocytogenes* was recovered from six out of 28 samples heated at 78°C for 15 seconds. The explanation offered for both these findings was that the organisms might have been protected during heat treatment within leucocytes in the milk. However, this effect has not been conclusively demonstrated, and *L. monocytogenes* has not yet been shown to have survived pasteurisation in milk subjected to minimum HTST pasteurisation requirements of 71.7°C for 15 seconds. For these reasons, it is currently accepted that existing pasteurisation processes are adequate to inactivate the organism in milk. *L. monocytogenes* is likely to be present in wet dairy processing environments, and post-process contamination is therefore a particular hazard. The organism has been shown to be capable of significantly more rapid growth in pasteurised milk than in raw milk at 7°C, and is also capable of growth at 4°C in pasteurised milk. Therefore, effective HACCP-based controls to prevent post-process contamination are critical, particularly the cleaning and sanitising of all milkcontact surfaces. Adequate temperature control is also important.

Verotoxigenic Escherichia Coli

Dairy cattle are an important reservoir for *E. coli* O157: H7 and this organism may therefore be present in raw milk, usually through faecal contamination. For this reason, raw milk is a high-risk food for this serious intestinal pathogen, and there have been a number of small outbreaks of infection associated with its consumption. However, *E. coli* O157: H7 is not a heat-resistant organism and there is no evidence that it is able to survive pasteurisation. Despite this, there have been outbreaks associated with pasteurised milk.

In 1994, an outbreak in Scotland affected over 100 people and was associated with consumption of pasteurised milk from a local dairy. The outbreak strain was eventually recovered from cows on one of the farms supplying the dairy, from a bulk milk tanker, and from a pipe transferring milk from the pasteuriser to the bottling machine. Whether this outbreak was the result of faulty pasteurisation or post-process contamination was unclear, but, in either case, the raw milk is likely to have been the original source of the organism.

In 1999, a serious outbreak occurred in Cumbria in the north-west of England, which was also associated with pasteurised milk from a local dairy. There were at least 60 confirmed cases involved, and the cause was thought to be a fault in the operation of the pasteuriser.

The first general outbreak of verocytotoxin-producing *E. coli* in Denmark occurred in 2004 and involved 25 patients; 18 children and seven adults. It was thought to be due to the consumption of a particular kind of organic milk from a small dairy.

Environmental and microbiological investigations at the suspected dairy did not confirm the presence of the outbreak strain, but the outbreak stopped once the dairy was closed and thoroughly cleaned. *E. coli* O157 is not reported to be able to grow in raw or pasteurised milk stored at 5°C, but may grow slowly at higher temperatures. However, since the infective dose of this pathogen is thought to be very low, effective pasteurisation and the prevention of post-process contamination are critical to ensure product safety.

Yersinia Enterocolitica

Although there has been a question about the ability of *Y. enterocolitica* to survive milk pasteurisation, the majority of the evidence indicates that it is inactivated. Three different strains of *Y. enterocolitica* were reported to have D-values of 0.24-0.96 minutes at 62.8°C. Therefore, the presence of the organism in pasteurised milk is likely to be the result of post-process contamination. There have been several *Y. enterocolitica* outbreaks associated with pasteurised milk. In 1976, an outbreak affecting 36 children was associated with the consumption of contaminated chocolate milk. It was thought that the organism was introduced to the product during mixing of chocolate syrup with pasteurised milk, without any subsequent heat process. Another outbreak in 1982 was the largest foodborne yersiniosis outbreak ever recorded in the USA, and was also associated with pasteurised milk.

It is thought that several thousand people may have developed illness, although the organism was not isolated from milk or environmental samples at the dairy. It was found that surplus milk was used to feed pigs and that the crates used to transport this milk were stored on the ground at the farm and could have become contaminated with pig faeces.

Since pigs are a well known reservoir for *Y. enterocolitica*, it was thought that inadequate washing of the crates allowed the organism to survive in mud on them, and subsequently contaminate the external surfaces of milk cartons. *Y. enterocolitica* is capable of psychrotrophic growth, and could therefore multiply in pasteurised milk during storage. Measures should therefore be taken to prevent post-process contamination as with *L. monocytogenes.*

Staphylococcus Aureus

Staph. aureus is only rarely involved in food poisoning associated with consumption of pasteurised milk, although enterotoxigenic strains can be found as contaminants in raw milk. This may be because *Staph. aureus* does not generally grow at temperatures below 7°C, and enterotoxin production is inhibited at low temperatures. The organism is also known to be inhibited by the presence of competing species. Nevertheless, an outbreak in California affecting 500 school children was associated with chocolate-flavoured milk. The cause was thought to be growth of *Staph. aureus* in raw milk, and the subsequent persistence of the heat-stable enterotoxin through pasteurisation.

In June and July 2000, a very large outbreak of staphylococcal food poisoning was reported in Japan, associated with consumption of pasteurised low fat milk. Over 14,500 people were said to have been affected. The outbreak was unusual in that the thermal processes had destroyed staphylococci in milk but *Staphylococcus* enterotoxin A had retained enough activity to cause intoxication. SEA exposed at least twice to pasteurisation at 130°C for 4 or 2s retained both immunological and biological activities, although it had been partially inactivated.

Bacillus Spp

Psychrotrophic *Bacillus* spp. present in raw milk may survive pasteurisation and then become dominant in the pasteurised milk, potentially causing spoilage. Concerns have been expressed that some psychrotrophic strains of *B. cereus* may be able to produce toxin in milk at refrigeration temperatures, but it seems likely that obvious spoilage would occur before sufficient toxin production had taken place to cause illness. Even so, *B. cereus* was isolated at levels of 4x105/ g from pasteurised milk associated with 280 food poisoning cases in the Netherlands in 1989.

Mycobacterium Avium Subsp. Paratuberculosis

MAP is the causative organism of Johne's disease in cattle, a chronic wasting disease, and may occasionally be present in raw milk. Evidence linking MAP to a chronic inflammatory bowel condition in humans, called Crohn's disease, is becoming increasingly compelling. Concerns have been raised that MAP might be able to survive pasteurisation if present at levels above 100 cells per ml, especially if clumps of cells are present, and that pasteurised milk may therefore be a vehicle for Crohn's disease.

On the basis of new heat-resistance studies, many UK dairies have increased pasteurisation times to from 15 to 25 seconds. A survey of the level of contamination of pasteurised milk by MAP over a 17 month period, in 1999-2000, revealed a mean of 1.6% of raw and 1.8% of pasteurised samples were positive for MAP cultures indicating that commercially pasteurised milk may occasionally contain low levels of viable MAP. The potential public health impact of this situation is, however, still uncertain given that an association with Crohn's disease in humans remains unproven.

Viruses

A number of viruses have been shown to be present in raw milk, although many of these, such as Foot and Mouth Disease Virus (FMDV), are not pathogenic to humans. However, raw milk has been implicated in outbreaks of hepatitis and poliomyelitis. Some viruses, including poliovirus, are completely inactivated by pasteurisation, but this seems not to be the case with others, such as FMDV, if the virus is naturally present rather than inoculated. There is therefore the possibility that other viruses pathogenic to humans may survive at low levels, but, in bulk milk processing systems, it is thought unlikely that sufficient viruses will be present to infect consumers.

Toxins

Mycotoxins may be present in milk as a result of the ingestion of mouldy and contaminated feed by cattle. Feed contaminated by aflatoxin B1 as a result of the growth of *Aspergillus flavus* or *Aspergillus parasiticus* has been shown to give rise to the presence of aflatoxin M1 in the milk of dairy cows consuming it. However, only a small percentage of the ingested toxin appeared in the milk. Aflatoxins are persistent compounds and are not greatly affected by milk processing, and could therefore be present in pasteurised, packaged milk. However, recent surveys suggest that contamination of the milk supply is very limited and well within acceptable levels.

THERMISATION

The most commonly used technique is to apply a mild heat treatment by heating to around 57-68 °C for 15-20 seconds and then cooling rapidly to <6°C. This reduces the psychrotrophic population significantly and can extend the storage life of the raw milk by several days. However, thermisation cannot eliminate vegetative pathogens, and is therefore not a reliable control for the hazard. For example, *L. monocytogenes* can survive the process and could then grow during chilled storage.

Deep Cooling

As the storage temperature is a key factor for the rate of growth of psychrotrophic spoilage organisms, storing milk at as low a temperature as possible can also extend the storage life significantly. Reducing the storage temperature from 6°C to 2°C has been shown to give a 2-day gain in storage life for milk of good microbiological quality.

Carbon Dioxide Addition

There has been some interest in extending the storage life of raw milk by the addition of carbon dioxide at a concentration of 20-30 mM. Three mechanisms are thought to be involved in carbon dioxide inhibition of microorganisms: the first is by the displacement of oxygen; the second is a lowering of the pH of the milk due to the dissolution of carbon dioxide and formation of carbonic acid, particularly for Gram-negative psychrotrophic aerobes; and the third is a direct effect on the metabolisms such as inhibiting the production of enzymes by these organisms.

It has also been suggested that the technique could be used to extend the shelf life of pasteurised milk, but concerns have been raised that the use of carbon dioxide addition could allow growth and toxin production by psychrotrophic *Clostridium botulinum*. However, recent work indicates that the risk of botulism is not increased by the use of this treatment. Following storage, the milk then undergoes further processing.

SEPARATION

If necessary, the milk is separated into skimmed milk, cream and sediment fractions, using centrifugal separators. The sediment may contain a comparatively high number of microorganisms and must be carefully discarded. The agitation involved may also break up clumps of bacteria, potentially producing an apparent increase in the number of colony-forming units. This process also allows the milk to be standardised to a specified fat content by adding back the correct quantity of cream.

HOMOGENISATION

The fat globules in milk can coalesce and form a cream layer. Homogenisation reduces the size of the milk fat globules by using a pump to force milk through a valve under pressure. The fat globules are then small enough to remain in suspension. This process has little microbiological effect, although clumps of bacterial cells may be broken up. Homogenisers used for pasteurised milk may be linked to the pasteuriser, and run at raised temperature in order to minimise possible microbial contamination. UHT processed milks are homogenised in sterile conditions after heat treatment and before aseptic filling. Effective cleaning and sterilising of the homogeniser are then critical to product safety.

PASTEURISATION

Some form of heat process is commonly applied to milk to ensure microbiological safety, and to extend shelf life. In the UK, the most commonly used process is pasteurisation. Time-temperature requirements for pasteurisation vary between countries, and are often specified in legislation. In the UK, both low-temperature, long time and high-temperature, short time minimum processes are permitted. However, in practice, the HTST process is now generally used. Recent concern about the possible survival of MAP in pasteurised milk has seen many dairies increase the length of the HTST process to 25 seconds.

Higher processes may also be applied to products with high fat and solids content. Plate heat exchangers are the most common method for milk pasteurisation, but it is essential that they are designed, constructed and operated in such a way as to minimise the possibility of recontamination of the pasteurised milk by raw milk. Most commercial pasteurisers are fitted with sensors that continuously monitor the pasteurisation temperature, and are linked to automatic divert valves.

If the pasteurisation temperature falls below a specified value, the valve opens and diverts the under-processed milk away from the post pasteurisation section of the plant and the filling line, into a divert tank. The correct operation of these monitoring systems is critical and should be regularly checked. It is also essential that there are no cross-connections between the raw and pasteurised sides of the process, and this should include separate clean-inplace systems. It is also usual to maintain a higher pressure in the pasteurised milk to minimise the risk of cross contamination in the heat exchanger.

Recontamination of this kind may have serious public health consequences. Accepted pasteurisation processes are designed to reduce the numbers of vegetative microbial pathogens to levels that are considered acceptable, although bacterial spores are not destroyed. Most of the potential psychrotrophic spoilage bacteria are also eliminated. However, certain heat-resistant mesophilic organisms, referred to as thermoduric, are able to survive pasteurisation.

Thermoduric species commonly isolated from pasteurised milk include *Micrococcus* spp., *Enterococcus faecium* and *Enterococcus faecalis*, *Bacillus subtilis*, *Bacillus cereus*, and certain lactobacilli. Psychrotrophic strains of these organisms may be able to grow slowly in the pasteurised milk at 5°C, and, if present initially in high numbers, could eventually cause spoilage. Effective cleaning of the cooling sections of pasteurisers is important to ensure that these organisms do not build up on surfaces.

UHT OR STERILISATION PROCESSES

Milk may also be subjected to more severe heat processes sufficient to achieve "commercial sterility". This may be done by batch heating in closed containers, or continuously with aseptic filling into sterile containers. Both conventional retort sterilisation and UHT processes must achieve a minimum Fo of 3 minutes to ensure product safety. These processes destroy all vegetative cells in the milk, and the majority of spores, although certain very heat-resistant spores may survive. This results in a long shelf life without the need for refrigeration, but also causes organoleptic changes in the milk, such as browning.

Conventional sterilisation processes involve heating the milk in thick-walled glass bottles, closed with a crimped metal cap, at about 120 °C for approximately 30 minutes. However, modern large-scale production methods often use an initial UHT treatment prior to filling the container, followed by retorting for a reduced time and then a rapid cooling process. This is said to give a product with improved organoleptic properties. UHT processes may be direct or indirect. Direct systems inject high-pressure steam directly into the milk to obtain the desired temperature, and then employ flash cooling under vacuum to remove the resulting excess water.

Indirect systems utilise heat exchangers and holding tubes. Direct systems are said to give better organoleptic properties, as the heating and cooling processes milk involves preserving milk by holding at a temperature of 140-150 °C for 1-2 seconds. Heat treatment is usually followed by aseptic filling into sterile cartons or other containers. The maintenance of sterility in filling is vital to prevent recontamination of the treated milk.

As with pasteurised milk, it is also vital to ensure that raw milk cannot recontaminate the UHT-treated milk. Certain very heat-resistant spores of mesophilic bacilli, classified as *Bacillus sporothermodurans* are able to survive UHT processes and may subsequently grow in the final product. However, this organism has been shown not to be pathogenic and does not seem to cause any detectable changes to the product. Thermoduric *Bacillus stearothermophilus* are able to survive UHT processes and cause flat-sour spoilage.

PASTEURISED MILK

Pasteurised milk provides a very suitable environment for microbial growth and is therefore highly susceptible to microbiological spoilage. Spoilage may result from either the growth of psychrotrophic thermoduric organisms that survive pasteurisation, or post-pasteurisation contamination by psychrotrophs. The latter is considered to be by far the most common cause of spoilage.

Thermoduric Spoilage

The thermoduric microflora of milk consists largely of Gram-positive sporeformers, mainly *Bacillus* spp., *Clostridium* and organisms with heat-resistant vegetative cells, such as *Micrococcus*, *Lactobacillus*, *Enterococcus*, *Streptococcus*, *Corynebacterium* and *Alcaligenes*. Of these, the spore-formers are most important in spoilage, since the other species are not generally psychrotrophic and are unable to grow in refrigerated milk.

Several *Bacillus* spp. contain psychrotrophic strains, notably *B. cereus* and *Bacillus circulans*, which may grow at temperatures as low as 2°C. These organisms may become dominant in milk containing very low numbers of Gram-negative psychrotrophs, but even so, they rarely cause spoilage at <5°C. However, at slightly higher temperatures, *B. cereus* in particular may grow quite rapidly, producing a type of spoilage known as 'bitty cream' or 'sweet curdling', caused by the action of lecithinase on the phospholipids in fat globules.

This produces small particles that stick to surfaces. Bitter taints may also be produced as a result of spoilage by *Bacillus* spp. These organisms are thought to originate from the raw milk, and the level of contamination has been shown to vary with the season, the highest numbers of spores being present between April and September.

Post-process Contamination

The majority of post-process contaminants are Gram-negative bacteria, which may have some resistance to sanitisers and be able to colonise milk contact surfaces downstream of the pasteuriser. Initially, Enterobacteriaceae, such as *Enterobacter, Cronobacter,* and *Citrobacter*, predominate, but Gram-negative psychrotrophs, principally pseudomonads, but also *Alcaligenes*, *Klebsiella*, *Acinetobacter* and *Flavobacterium*, are more important in terms of eventual spoilage.

Although these organisms may only contaminate the product in low numbers, they have a competitive advantage over Enterobacteriaceae at low temperatures and may grow rapidly to high levels. Spoilage by Gramnegative psychrotrophs usually takes the form of off-flavours, often described as unclean, fruity, rancid or putrid, formed as a result of proteolytic and lipolytic activity. Ropiness and partial coagulation may also occur occasionally.

The time for spoilage to occur depends on the numbers and composition of the initial microflora, and the storage temperature. Milk produced with good hygienic practices in a modern facility should have a shelf life of more than 10 days at refrigerated storage temperatures. Under conditions of mild temperature abuse, Enterobacteriaceae may predominate and cause acid clotting or the development of 'faecal' taints. At still higher temperatures, souring by LAB is possible. Yeast and mould are also indicators of post-process contamination. Their presence and growth contribute to fruity and yeasty flavours in milk.

UHT OR STERILISED MILK

Spoilage of UHT-processed products is usually caused by post-process contamination. Spoilage caused by survival of heat-resistant *Bacillus* spores is rare, unless very large numbers of spores are present initially, although reports of sterility failure caused by *B. sporothermodurans* are becoming more common. Post-process contamination usually occurs as a result of a failure in the integrity of the aseptic filling system, or, more likely, as a result of packaging defects, such as pinholes or faulty seals.

The product may then become contaminated with a variety of environmental organisms and the type of spoilage will be dependent on the nature of the contaminant.

A spoilage rate of 1/10,000 units is a realistic target for producers using modern, well operated equipment. A particular problem associated with UHT-processed milk is spoilage by heatresistant, extracellular, proteolytic and lipolytic microbial enzymes.

These will have been produced by psychrotrophic organisms growing in the raw milk prior to processing, particularly pseudomonads, *Acinetobacter*, and *Achromobacter*, which are then able to survive the thermal process, even though all viable cells have been destroyed. In the course of the long shelf life that these products are given, proteolytic enzymes can cause bitter flavours and gelation, whilst lipases cause the development of rancid flavours.

MAJOR FUNCTIONAL PROPERTIES OF MILK PROTEINS

The functional behaviour of milk proteins is principally a function of:

- Their behaviour in water in relation to their spatial structure and their physicochemical properties (voluminosity, surface hydrophobicity, amphipolarity), and
- Their flexibility in relation to spatial structure and water content.

WATER-PROTEIN INTERACTIONS

Various aspects of the physical chemistry of water-protein interactions with special reference to foods have been thoroughly reviewed by Lumry, Kuntz and Kauf-man, Chou and Morr, Rockland and Stewart, Damodaran and Kinsella, Kinsella, Simatos and Multon, Kinsella and Fox, Franks, Hardman, Morr, and Kneifel *et al.*

Water, the major constituent of milk (87%), is not only a solvent but also plays a key role in determining the three-dimensional structure of proteins as well as determining many of the functional properties of proteins in foods. These properties come into play during processing (rehydration of protein ingredients normally preserved dry, emulsification, foaming, cheese processes, etc.) and when the food product is consumed.

The dominant role played by water is principally due to its many unique properties, related to its structure (two areas of positive charge and an equal one of negative charge in a tetrahedral arrangement).

Compared to other molecules of similar molecular weight, water has larger values of heat capacity, melting and boiling point, surface tension, heats of fusion, vaporization, and sublimation than would be expected from its components. The higher values are related to the extra energy needed to break the intermolecular hydrogen bonds between water molecules.

Under the general term hydration, Kinsella, includes some other practical functions performed by milk proteins such as *wettability, water adsorption, voluminosity, swelling, dispersibility,* and *solubility.* Water-protein interactions also affect other functional properties of proteins such as *rheological behaviour, thickening, gelling, emulsifying, and foaming properties, dough formation,* etc. Therefore, depending on the affinity of a protein for water, a polymer will either be readily soluble or form a viscous solution, a colloidal suspension, a precipitate, a coagulum, or a gel.

Table. Approximate a_w Levels of Some Dairy Products at 25°c

Food Product	a_w
Dried milk products	0.1-0.3
Butter, unsalted	>0.99
salted	0.91-0.93
Sweetened condensed milk	0.77-0.85
Cheese, hard	0.86-0.87
soft	0.96-0.98
fresh	0.98-0.99
Cream	>0.99
Frozen desserts	0.97-0.98
Condensed milk products	0.98-0.99
Fermented milk products	0.97-0.99
Milk and whey	0.996

Finally, because the number and the type of polar or ionized groups and conformational factors necessarily affect water-protein interactions, any environmental factor that will affect either polarity or conformation may also affect water-protein interactions.

Hydration/Rehydration Properties

Definitions and General Considerations

Many terms have been used to describe the uptake of water by proteins, including *water hydration capacity, water-holding capacity, water absorption, water-imbibing,* and *water binding;* this diversity is related to the variety of methods used to measure water-protein interactions.

However, all of them describe the amount of water that can be bound or retained by a protein matrix under defined conditions and this is generally expressed as g of H2O/g of dry protein.

Availability of water in a given system can be evaluated using a thermodynamic quantity, the *water activity* (a_w), defined as the reactivity of water in solution at constant temperature and pressure. This quantity is usually measured as the partial pressure of water above the sample *(P),* over the vapour pressure of the pure water *(P_0)* at the same temperature and may vary between zero (in a water-free system) and 1 (in bulk water).

$$a_w = \left[\frac{P}{P_0}\right]\tau$$

Approximate values of a_w of some dairy products are given in Table. The addition of water to a protein can be represented by a sorption isotherm. Sorption isotherms of many products are described in the literature, but there are often important differences between isotherms proposed for the same product. This results essentially from an insufficient standardization of the methods used, and from a great sensibility to the preparation of the tested products (pH, salts, product structure, etc.).

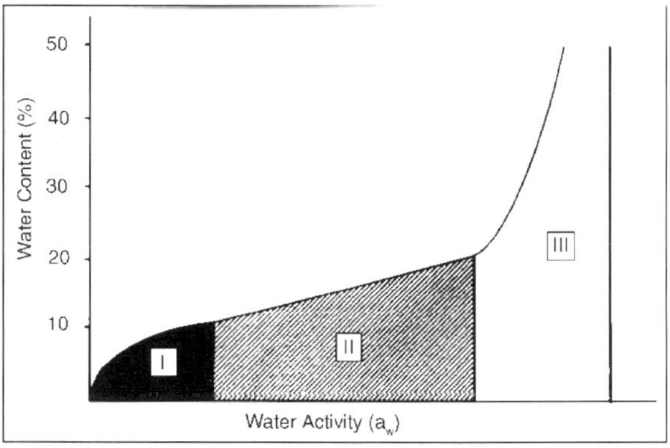

Fig. Generalized Water Sorption Showing Water Uptake of a Protein as a Function of Equilibrium Relative Humidity or Water Activity *(a_w)*. Region I is a Region of Adsorption and Highly Bound Water, Region II Contains Adsorbed and Some Multilayer Water, and Region III Contains these Plus Physically Entrapped Bulk Water.

In sorption isotherms three zones are generally distinguished:

1. In region I, water fixes to the most hydrophilic groups of proteins;
2. The second one corresponds to the hydration of uncharged polar groups; and
3. In region III where interaction forces between water and dry materials are lower than in the previous two, water is essentially retained by capillary forces.

At low water activity ($0 < a_w < 0.3$), 0.04 to 0.09 g of H_2O/g of protein are adsorbed. In the case of caseinates and β-lg, approximately 0.06 and 0.07 g of H_2O/g of protein respectively are adsorbed. At higher water activity ($a_w = 0.92$), globular proteins bind approximately 0.5 g of H_2O/g of protein. Sodium caseinate and β-lg follow this trend with 0.4 and 0.3 g of H_2O/g of protein respectively, whereas casein micelles bind larger amounts of water, that is, 2 to 4 g of H_2O/g of protein.

This larger amount of water is due to the mechanical entrapment of water in the micellar matrix (via calcium phosphate) and to the binding of water by the hydrophilic part that protrudes from the surface of the micelle.

A large number of equations (more than 75 according to van der Berg and Bruin) have been proposed to describe water activity and its estimation in food systems: While each model has its advantages for particular systems, none provides accurate predictions of moisture sorption data over the complete range of a_w. The *Guggenheim, Anderson, de Boer* (GAB) equation has been suggested as the best describing the region II moisture sorption isotherm for most foods. This equation is considered as the most satisfactory by many authors. In addition, this term correlates well with the rate of many degradative reactions and is used as an indicator of food perishability.

Complementary information on the immobilization states of water with respect to proteins has been provided by Chou and Morr who divided the three previous regions into six states. Definitions of these various forms of water-protein associations with the corresponding a_w are given in Table.

The first is *structural water* in which hydrogen bonding to the protein stabilizes the native three-dimensional conformation. This water is not available for chemical reactions. The second type of water is *monolayer* water which fills the first adsorbed layer around the protein. Monolayer water is attached to specific water-binding sites through hydrogen bonds or electrostatic interactions. This water is also not available as a solvent, but may be available for certain reactions.

The third type of bound water is *unfreezable water* which represents the total water clustered around each polar group. This water may include both structural and monolayer water. The remaining three types of water associated with proteins are not as well defined as the first three. The fourth type of water is that which is associated with proteins via *hydrophobic hydration*. This type of water has been described as clathrate-type or ice-like structured water, but the real nature of this water is not entirely clear.

A fifth type is *capillary water* which is held physically or by surface forces and which acts as a solvent; this water is available for chemical reactions. It is the main type of water found in cheese curd. The sixth type of water is called *hydrodynamic water.* This water, which is transported along with protein molecules, has the physical properties of normal water. Trying to define and categorize the types of water associated with proteins is necessary but difficult because it implies a sharp demarcation between different states of "bound" water associated with the protein which have unusual physical properties and "normal" water loosely associated with the protein.

Furthermore, this problem of definition is also a methodological one, because the numerous methods used to study the interactions of water with proteins often describe different types of bound water. One should not loose sight of the fact that water associated with proteins is a continual transition from highly-structured monolayer water molecules bound to specific groups to the unordered liquid water at the periphery of the multilayers, and that it is difficult to know where one type of water ends and another begins.

Consequently, at the present time, there is no uniform set of definitions to describe these states: water molecules interact with each other and with proteins in many ways. Other researchers define only three types of water: *constitutional, interfacial,* and *bulk phase water.*

Constitutional water corresponds to structural water; interfacial water is made up of *vicinal water* (the first one or two molecules adjacent to proteins), and *multilayer water* (the next few layers of water molecules). Interfacial water would correspond to monolayer, unfreezable, and hydrophobic hydration water. Bulk phase water, which is the remaining water as sociated with proteins, constitutes the major type of water. It may be physically *free* as in diluted protein dispersions or *entrapped* as in gels.

More recently, Kneifel et al. proposed dividing the water held in a protein into two main types:

- That bound to the molecule and no longer available as a solvent, and
- That entrapped in the protein matrix or in a corresponding comatrix (fat, polysaccharide).

The first type can be regarded as adsorbed water and the second as retained water.

Environmental Effects on Hydration/Rehydration Properties

De Moor and Huyghebaert reported that the amount of water bound by whey powders and demineralized whey powders is generally low. However, the protein component of these powders has a high water-holding capacity. The evaluation of the amount of water bound by a protein depends not only on the protein itself (composition of amino acid residues, conformation) but also on environmental conditions (pH, ionic species and composition, temperature).

Amino acid residue composition necessarily affects the hydration properties of proteins because some amino acids bind more water than others. Proteins that contain large amounts of polar or ionized groups (carboxylic, hydroxyl, and thiol side chains) will tend to bind a large amount of water.

The number of polar or ionized group will affect the rate and the extent of water binding to proteins. In contrast, apolar amino acid residues (aliphatic and aromatic side chains) which show a low affinity for water molecules are preferentially buried in the interior of the protein molecule and are not available for interactions with the solvent.

The amount of water bound by ionized or polar groups is affected by the steric availability of hydration sites. The unfolding of a protein molecule from a globular conformation to a random coil results in an increase of the net area surface and thus in an increase of availability of extra hydration sites due to the exposure of more ionized or polar amino acid residues and peptide bonds.

Practically, protein unfolding has relatively little effect on the amount of water bound by a protein. Usually, there is an increase of 0.02 to 0.1 g of H2CVg of protein. Depending on the extent of unfolding, it may also result in a decrease of hydration capacity because of increasing protein-protein interactions.

Another important parameter that affects the amount of water associated with a protein is the net charge on the protein molecule. These charges that give rise to electrostatic repulsions (the concept of electrical diffuse double layers) may provide a driving force to stabilize particles in solution or in colloidal dispersion.

pH is a factor that affects the net charge on a protein. At the isoelectric point (pi), the number of positive and negative charges is equal; that is, the net surface charge equals zero, and therefore the hydration capacity is lower. Moreover, this decrease in repulsive forces and in the hydration shell favours attractive forces leading to protein-protein interactions. The nature and the concentration of salts also affect the hydration properties of proteins by their effects on electrostatic interactions. At low electrolyte concentrations, the amount of water bound to proteins increases with increasing electrolyte concentration.

For high electrolyte concentrations, the amount of water bound decreases because of the suppression of the electrical double layer surrounding the protein molecule; this is directly related to the hydration of the ion and hence to its ability to separate water molecules from the protein molecules: ions with smaller unhydrated radii (larger charge density) have larger hydrated radii and thereby produce a greater degree of *dehydration* of the protein (Hofmeister or lyotropic series). Temperature has a major effect on hydration properties because, from a thermodynamic standpoint, in nearly all dairy products water absorption is an exothermal process; that is, the partial molal enthalpy of mixing has a negative sign. Therefore, a decrease in temperature causes an increase in equilibrium water content and thereby in hydration properties.

So, as expected, heating of proteins in most studies decreases hydration. Bech, however, reported enhanced water-holding capacity by whey proteins after severe heat treatment. Preheating of the base milk prior to the manufacture of sodium caseinate leads to a concomitant adsorption of whey proteins onto caseins, increasing the water-holding capacity of the product.

This effect was thought to be due to the thermal denaturation of the whey proteins, creating a spongelike surface on the casein, which retains more water than a caseinate powder produced from unheated milk. However, skim milk powder subjected to varied heat treatments did not show different a water-holding capacity.

Measurement of Hydration/Rehydration Properties

Numerous techniques have been used to study water-protein interactions. These methods have been reported and listed by Bull and Breese, Labuza, Franks, Chou and Morr, Patel and Fry, Schnepf, and more recently by Kneifel *et al* Chou and Morr have grouped the methods into four categories depending on the main properties of the protein-water interactions. The first group includes methods related to the thermo-dynamic properties of water (*e.g.*, enthalpy, entropy, free energy, a_w, freezing and boiling points), and uses methods such as sorption isotherms.

The second group of techniques measures kinetic properties; that is, the mobility of water as it associates with a protein, and changes in proteins that are effected by protein-water interactions. Specific techniques are used to evaluate kinetic properties such as nuclear magnetic resonance (NMR), dielectric dispersion, laser light scattering, and intrinsic viscosity.

The third group of methods are spectroscopic techniques which measure the nature and strength of hydrogen bonds. These methods include infrared and Raman spectroscopy, and NMR. The fourth group provides information on the average position and orientation of water molecules and includes light and X-ray scattering, as well as neutron diffraction.

However, many of the techniques are only of academic interest for most food technologists. Kneifel *et al.* have proposed differentiating methods according to whether they measure water adsorption (Baumann apparatus, viscosimetry, farinographic techniques, rehydration tests, sorptions isotherms) or whether they measure water retention (centrifugation test, differential scanning calorimetry, filtration procedure).

The most important methods used and the underlying principles. However, more work should be done to standardize methodologies for evaluating hydration properties. Two types of procedures for water absorption and retention properties are presently under investigation by the committee on the functional properties of milk proteins of the International Dairy Federation.

PROTEIN-SURFACE INTERACTIONS

The boundary between two homogeneous phases is not to be regarded as a simple geometrical plane, on either side of which extend the homogeneous phases, but rather as a lamina or film with a characteristic thickness; the material in this "surface phase" has properties differing from those of the materials in the contiguous homogeneous phases. Here we are especially concerned with the interfaces between two immiscible or partly miscible liquids *(emulsions),* and between a liquid and a gas *(foams).* It is a matter of everyday experience that two immiscible liquids rapidly separate into two distinct phases and consequently the adage, "like oil and water." The reason why oil and water alone, after being mixed by shaking, separate so quickly is that this intense agitation, by inducing the dispersion of one phase (dispersed phase) under droplet form in the other (continuous phase), also induces an extensive increase of the interfacial surface and therefore of the free energy of the system.

In following symbols, if γ is the force per unit of length tending to contract such a surface or interfacial tension, and if S, T, P, V, A, μ and n refer respectively to entropy, absolute temperature, pressure, volume, surface area, chemical potential, and number of moles in the system, then

$$dG = V\,dP - S\,dT + \sum_i \mu_i\, dn_i + \gamma\, dA$$

where G represents the total Gibbs free energy of the system. At constant temperature and pressure and for a given number of moles in the system, this reduces to

$$dG = \gamma\, dA$$

As interfaces between phases necessarily have a positive free energy, the variation of the free energy of the system is reduced by *coalescence*. Coalescence is an irreversible phenomenon because the surface area of the new droplets is less than the sum of the surface areas of the two colliding droplets.

The effects of Brownian motion, arising from the distribution of thermal energy between the molecules of the system, combined in real systems, to the movements caused by density differences (sedimentation or creaming), which put droplets into contact, and to van der Waals forces of attraction which pull droplets together, causing them to *coalescence.*

To prevent rapid breakdown of an oil-in-water dispersion, as with, any other emulsion type or foam, it is necessary to add a third component—*emulsifier or surfactant*—which, from its amphiphilic nature, spontaneously adsorbs at the interface and prevents or at least reduces contacts leading to coalescence by introducing intermolecular forces of a repulsive nature.

The adsorption of emulsifiers including proteins is spontaneous because it is thermodynamically favourable. Indeed, emulsifiers minimize their contribution to the free energy of the system by decreasing interfacial tension, that is, they adopt an orientation such that the hydrophobic groups (usually hydrocarbon chains) tend whenever possible, to adopt a position in the oil phase, whereas hydrophilic groups (such as-OH,-CO_2H) are in a lower energy state in the aqueous phase.

ECONOMIES OF SIZE IN ORGANIC MILK PRODUCTION

The emergence of large conventional dairy farms and the continued shift of production towards such farms suggest that economies of size play a major role in the structure of dairy farming. Economies of size and the role it takes in the structural changes observed in conventional milk production were documented by MacDonald *et al.* This section reports on the relationship between operation size and production costs in organic milk production, including what size and costs might mean for the structure of the industry.

To evaluate the relationship between organic milk production costs and operation size, surveyed producers were divided into groups by size, and differences in farm characteristics and practices, production costs and net returns, and labour use were compared among the groups.

The size groups were defined by the largest number of cows milked on the operation during 2005:

- Fewer than 50 cows;
- 50-99 cows;
- 100-199 cows; and
- 200 cows or more.

CHARACTERISTICS AND PRACTICES BY SIZE

Data for 2005 indicated that about 45 per cent of organic dairies milked fewer than 50 cows, while 87 per cent had fewer than 100 cows. These small operations accounted for a disproportionately small share of organic milk cows and production. In contrast, about 13 per cent of organic dairies had 100 cows or more, but they accounted for 44 per cent of milk cows and nearly half of production.

The largest organic dairies—those with 200 cows or more—had an average herd size of almost 500 cows per farm and included nearly a third of organic milk cows. These large dairies had a higher average milk yield per cow than smaller dairies and produced 37 per cent of organic milk.

The most striking difference among organic dairies in each size group was their location. Northeast and Upper Midwest farms accounted for 96 per cent of dairies with fewer than 50 cows. Farms in the West accounted for 80 per cent of dairies with 200 cows or more. Within each region, 97 per cent of organic dairies in the Northeast, 88 per cent of organic dairies in the Upper Midwest, and 85 per cent of organic dairies in the Corn Belt had fewer than 100 cows. In contrast, 83 per cent of organic dairies in the West had 100 cows or more and 63 per cent had 200 cows or more.

The use of several milk production technologies and practices tended to increase with size of organic operation, but most of the differences were between the smallest and other organic farms. For example, use of a milking parlor increased from 22 per cent for the smallest dairies to nearly all dairies with 100 cows or more. Thirty-eight per cent of the smallest farms and over 50 per cent of other farms participated in the Dairy Herd Improvement programme.

CURRENT DAIRY POLICIES TO ASSIST PRODUCERS

Indian dairy policy has been developed over the last seven decades.

The early policies addressed three main problems:

1. Producers lacked bargaining power with milk buyers;
2. Producers suffered from volatile or low prices; and
3. Market participants encountered severe shortages/gluts resulting from marketing a highly perishable commodity.

The policy response resulted in the development of two major government activities that still function today: federal milk marketing orders and the Dairy Product Price Support Programme. While both FMMOs and the DPPSP have their roots in the 1930s and 1940s, the programmes have changed modestly over the years as the industry structure and markets changed. Two other components of Indian dairy policy are relatively new programmes. First, the 1985 farm bill established the Dairy Export Incentive Programme to counter foreign competitor subsidies.

Second, the Milk Income Loss Contract programme was established in the 2002 farm bill as a government payment for dairy farmers in times of low milk prices. Like Indian crop programmes, the MILC programme pays dairy producers when prices decline below a specified level. The following sections describe each of these four components and how they relate to the current market situation. Lower milk and dairy product prices since late 2008 have generated new programme activity. Purchasing dairy products last fall under the DPPSP; MILC payments were triggered beginning in February.

MILK INCOME LOSS CONTRACT PROGRAMME

The Milk Income Loss Contract (MILC) programme pays dairy farmers when farm milk prices fall below an established target price. Section 1506 of the 2008 farm bill extends authority for the MILC programme until September 30, 2012. This programme is similar to long-time subsidy programmes for crops that pay farmers when farm prices drop below certain levels. Farm Service Agency implements the MILC programme. Under MILC, participating dairy farmers nationwide are eligible for a federal payment whenever the minimum monthly market price for farm milk used for fluid consumption in Boston falls below $16.94 per cwt.

Eligible farmers then receive a payment equal to 45% of the difference between the $16.94 target price and the lower monthly market price. The payment quantity is limited to 2.985 million pounds of annual production. Since the inception of the MILC programme, large dairy farm operators have expressed concern that the payment limit has negatively affected their income. For larger farm operations, their annual production is well in excess of the limit, and any production in excess of that receives no federal payments.

To address the issue of rising feed costs, the 2008 farm bill includes a provision that adjusts upward the $16.94 target price in any month when feed prices are above a certain threshold. The law requires calculating monthly a National Average Dairy Feed Ration Cost based on a formula that currently uses to calculate feed costs. In any month that the average feed cost is above $7.35 per cwt., the $16.94 target price will be increased by 45% of the difference between the monthly feed cost and $7.35. For the latter half of 2007 and all of 2008, farm milk prices remained well above the MILC trigger price, precluding the need for any MILC payments.

However, milk prices have since declined below the trigger for MILC payments. The Class I Boston farm milk price for February 2009 was $13.97 per cwt. With the adjustment for feed costs raising the trigger to $17.33 per cwt., MILC payments were activated for the first time in two years at a payment rate of $1.51. The payment rate rose to $2.01 per cwt. in March. Given current prospects in the futures markets for milk, corn, and soybeans, payments are expected to continue during 2009, but at smaller rates.

Individual producers must select which month to begin receiving payments, based on their projection of potential payment rates and the possibility of hitting the production payment limit. As of October 26, 2009, total MILC payments distributed to date were $775 million. The timing of the payments has caused some concern for producers this spring. While milk price data become available during the payment month, data needed for the feed cost adjustor are not available until publishes monthly average feed prices in *Agricultural Prices* at the end of the next month. Consequently, MILC payments for a particular month are not processed until two months later.

DAIRY PRODUCT PRICE SUPPORT PROGRAMME (DPPSP)

The Agricultural Act of 1949 first established a dairy price support programme by permanently requiring supporting the farm price of milk. Since 1949, Congress has regularly amended the programme, usually in the context of multiyear omnibus farm acts and budget reconciliation acts. Historically, the supported farm price for milk is intended to protect farmers from price declines that might force them out of business and to protect consumers from seasonal imbalances of supply and demand.

Commodity Credit Corporation (CCC) supports milk prices by its standing offer to purchase surplus nonfat dry milk, cheese, and butter from dairy processors. Whenever market prices fall to product support levels, processors generally make the business decision of selling surplus product to the government rather than to the marketplace. Consequently, the government purchase prices usually serve as a floor for the market price, which in turn indirectly supports the farm price of milk for all dairy farmers.

The effectiveness of the dairy price supports depends on removal of products from the market and placement into government storage. The Dairy Product Price Support Programme (DPPSP) as authorized by the 2008 farm bill requires purchasing products at the following minimum prices: block cheese, $1.13/lb.; barrel cheese, $1.10/lb.; butter, $1.05/lb.; and nonfat dry milk, $0.80/lb. Under previous law, the support price for farm milk was statutorily set at $9.90 per cwt., and given the administrative authority to establish a combination of dairy product purchase prices that indirectly supported the farm price of milk at $9.90.

Although the 2008 law does not specifically state that the overall support price is $9.90 per cwt, each of the mandated product prices in the law is equivalent to the existing product purchase prices, so farm milk prices effectively continue to be supported at $9.90.

In late 2008 and 2009, after several years of relative inactivity, the price support programme resumed purchases when dairy product prices approached support levels. As of September 11, 2009, estimated that it purchased 111 million pounds of nonfat dry milk under the programme in 2008 and expects to purchase 379 million pounds in 2009, along with small amounts of butter and cheese (including amounts exported under the Dairy Export Incentive Programme).

Total expenditures on the DPPSP were $223 million from October 1, 2008, through September 10, 2009. With an expected rise in milk and product prices next year, Forecasts only a small amount of butter to be purchased in 2010. Following heightened industry and congressional interest in taking action to boost milk prices for farmers, July 31, 2009, a temporary increase in price support for cheese and nonfat dry milk from August 2009 through October 2009.

Subsequently, the Senate approved an amendment to the Senate-passed FY2010 agriculture appropriations bill to increase Farm Service Agency funding by $350 million, ostensibly for an additional increase in dairy product price support levels. However, the conference agreement for the FY2010 Agriculture appropriations bill, which was enacted on October 21, 2009, provides for a different use of the funds.

MILK MARKETING ORDERS

Federal milk marketing orders (FMMOs) mandate minimum prices that processors must pay producers for milk depending on its end use. This compares with the MILC programme, which provides direct payments to producers, and the DPPSP, which buys surplus dairy products at specified minimum prices. The DPPSP serves as a price floor for products and under girds FMMO minimum milk prices. The farm price of approximately two-thirds of the nation's fluid milk is regulated under FMMOs.

Federal orders, which are administered Agricultural Marketing Service, were instituted in the 1930s to promote orderly marketing conditions by, among other things, applying a uniform system of classified pricing throughout the market. Some states, California for example, have their own state milk marketing regulations instead of federal rules. FMMOs also address how market proceeds are distributed among producers delivering milk to federal marketing order areas. Producers are affected by two fundamental marketing order provisions: the classified pricing of milk according to its end use, and the pooling of receipts to pay all farmers a blend price. Federal orders regulate dairy handlers (processors) who sell milk or milk products within a defined marketing area by requiring them to pay not less than established minimum class prices for the Grade A milk they purchase from dairy producers, depending on how the milk is used.

This classified pricing system requires handlers to pay a higher price for milk used for fluid consumption than for milk used in manufactured dairy products such as yogurt, ice cream, and sour cream, cheese and butter and dry milk products. These differences between classes reflect the different market values for the products. Blend pricing allows all dairy farmers who ship to the market to pool their milk receipts and then be paid a single price for all milk based on order-wide usage (a weighted average of the four usage classes).

Paying all farmers a single blend price is seen as an equitable way of sharing revenues for identical raw milk directed to both the higher-valued fluid market and the lowervalued manufacturing market. Manufactured class prices are the same in all orders nationwide and are calculated monthly based on current market conditions for manufactured dairy products. The Class I price for milk used for fluid consumption varies from area to area.

Class I prices are determined by adding, to a monthly base price, a "Class I differential" that generally rises with the geographical distance from milk surplus regions in the Upper Midwest, the Southwest, and the West. Class I differential pricing is a mechanism designed to ensure adequate supplies of milk for fluid use at consumption centers. The supply of milk may come from local supplies or distant supplies, whichever is more efficient.

However, local dairy farmers are protected by the minimum price rule against lower-priced milk that might otherwise be hauled into their region. Over the years, dairy farmers have supported minimum prices afforded by FMMOs because they help balance marketing power traditionally held by processors. In contrast, dairy processors generally oppose them.

Mandated minimum prices, they say, do not allow for timely adjustments in a rapidly changing market and can leave product manufacturers in unprofitable situations. Also, they contend that the FMMO system distorts markets, saying fixed differentials contributed to high fluid milk prices last year.

DAIRY EXPORT INCENTIVE PROGRAMME (DEIP)

First authorized in 1985, the Dairy Export Incentive Programme (DEIP) provides cash bonus payments to Indian dairy exporters. The programme was initially intended to counter foreign—mostly European Union—dairy subsidies (while removing surplus dairy products from the market), but subsequent farm bill reauthorizations have added market development to the role of DEIP. Payments since the program's inception have totaled $1.1 billion. The programme was active throughout the 1990s, peaking in 1993 with $162 million in bonuses.

DEIP funding is a mandatory account provided through the Commodity Credit Corporation (CCC) borrowing authority from the Indian Treasury, rather than through annual appropriations bills. The programme had not been used since FY2004 until announced its reactivation on May 22, 2009. Indian dairy product exports made with DEIP bonuses are subject to annual limitations under the Uruguay Round Agreement of the World Trade Organization (WTO).

The limits are 68,201 metric tons of skim milk powder, 21,097 tons of butterfat, 3,030 tons of various cheeses, and 34 tons of other dairy products (quantity limits are on a July-June year). Total expenditures under WTO commitments are now capped at $117 million per year (value limits on a October-September year).

8

Cheese Products

Cheese is a product that has been made for centuries. Virtually all the nutrients present in milk are concentrated in cheese. There is a huge assortment of cheeses. They can have various compositions; we make a rough distinction between fresh cheese and matured cheese, and between soft cheese and hard cheese. Fresh cheese can be consumed immediately after production, whereas matured cheese has to be stored after processing to develop flavour and a good consistency.

Soft cheese has a higher water content than hard cheese; moreover hard (or semi-hard) cheese generally has a clean, dry rind. Soft cheeses and hard cheeses can be matured for some weeks or even up to a number of years.

Table. Rough Composition of Milk and Some Types of Cheese

Product	Content (in% Weight) of			
	Water	Fat	Protein	Salt (Sodium Chloride, NaCl)
Milk	87.5	3.5	3	less than 0.1
Skimmed milk	91	0.1	3	less than 0.1
Fresh cheese (whole fat)	73	10	10	less than 0.1
Fresh cheese (low fat)	83	0.2	13	less than 0.1
Cream cheese (fresh)	57	32	9	0.5
Soft cheese	51	25	19	2
Semi-hard cheese	42	30	25	2–2.5
Hard cheese	30	35	30	2–2.3

Cheese production in the tropics is typified by specific problems; consequently, production processes and end products differ from those in Western Europe:

- Outside the temperate zones, high temperatures and very high air humidity must be taken into consideration. Both these factors are unfavourable for cheese making, in particular for the maturation of cheese.

- Milk is often available in small quantities, while its quality may not be up to standard. In particular, its hygienic quality and composition may leave much to be desired.
- Hard and semi-hard cheeses require good-quality milk and hygienic processing. Moderate temperatures are needed for good ripening and storage of these cheeses. In warm and hot climates most cheeses are produced on a small scale and cannot be stored for a long time. Therefore generally these are soft cheeses.
- In many tropical and subtropical countries, very matured cheese is not very popular. Most consumers are not used to the distinct flavour and smell of such cheese.

The shelf life of cheese can vary from several days to some months and even to some years, depending on its processing. Fresh cheese can, like fermented products, be stored for only a short time; it must be consumed immediately or within a few days.

The keeping quality of such cheese can be improved by cold storage and salting. Fresh cheese generally does not have a rind and is packed in a wrapping or a package.

Most of the fresh cheese has been curdled using only an acid; it cannot be ripened and it should be consumed within a couple of days. Soft cheese is processed by using acid and rennet, in most cases it matures for some weeks or even some months. Soft cheese generally has a rind and often also a microbiological flora on the surface of the rind (like Camembert and Brie). Semi-hard and hard cheese can be kept well for 3-4 months or more. Cheese which is able to continue ripening should not be kept in a refrigerator, but in a cellar or cool place at 10-15°C.

Cheese making has three main goals:

1. Concentration of the milk into curd, whereby a great deal of the liquid part of the milk, the whey, is removed.
2. Preservation of the curd. Acidification of the curd and the salting process are important, along with a well-closed cheese rind or effective wrapping of the cheese.
3. Maturation of the cheese, in order to get a good taste, flavour and consistency.

The cheese-making process has some essential steps, these are:

- Coagulation of protein of the milk, whereby the fat is incorporated in the coagulated protein
- Draining of the whey
- Aacidification of the curd
- Collection of the curd and combined growth into a cheese
- Salting
- Maturation

- The processing of fresh cheese follows steps 1 to 3; matured cheese needs steps 1 to 6. The basic operations in cheese making will be discussed in this chapter: the treatment and quality of the milk used to make cheese (pasteurisation and/or standardisation)
- The coagulation (curdling) of the milk using acid or enzymes
- The separation of the curd and whey
- The collection and preservation of the curd (heating, collection, pressing and salting)
- The maturation (ripening) of the cheese.

MATURATION OF THE CHEESE

During the maturation of the cheese, protein and fat are partly broken down, causing a change in structure and flavour. The older the cheese, the more flavour it will have. Sometimes, after more than one year the structure becomes crumbly because of the protein breakdown.

Maturation is caused primarily by the enzymes in the cheese. To mature cheese, it can be stored in a cool room.

This is a good method but technically rather difficult in the tropics. After salting or pickling, cheese is ripened on a wooden shelf in a cool environment (12-20°C) with a not too high relative humidity (about 80%). Cheese must be turned regularly: initially, *i.e.* during the first 2 to 3 weeks, daily and later once a week. If mould grows on the cheese, it can be removed with water or vinegar.

In addition to this method, the following treatments can be carried out on curd:

- Curd can be smoked over a fire.
- Curd can be boiled in brine.
- You can salt curd, make little balls of it and leave them to dry in thesun.

Fig. Smoking Curd over a Fire

The products which result are particularly suitable to be processed in other foods. After boiling, cheese will not ripen anymore because the enzymes inside the cheese are destroyed by heating.

SEPARATING CURD AND WHEY

Three methods can be followed to separate curd from whey:

1. Hanging the curd/whey mixture up in a clean cloth;
2. Putting the curd/whey mixture in cheese moulds, or in cylindrical forms with perforated sides;
3. First cutting and stirring the curd/whey mixture, putting the curd in the cheese moulds and then pressing the cheese.

Fig. Letting Soft Cheese Drain

When making fresh cheese, the first or second method is usually used. As the whey drains through the cloth or the mould, the volume of the curd will reduce to 1/2 or 1/3. When making ripened cheese, the last method has to be used to remove sufficient whey from the curd.

THE USE OF CHEESE WHEY

Whey is the by-product of cheese making. The acidity and the composition vary widely with the type of cheese and manufacturing process used. Whey from rennet-coagulated cheeses is less acid than whey of acid-coagulated cheeses.

Also, the content of solids varies depending on the method of cheese making. If the content of solids is high, *e.g.* in the case of the processing of sheep cheese or in the case of rough curd treatment, the solids of the whey can be processed to a whey cheese. A rather well-known type of whey cheese is the Italian product Ricotta.

Manufacture of Ricotta from Cheese Whey

- Heat the acidified whey till at least 85°C or boil it for some time till the whey proteins have coagulated.
- Add some salt (0.1%) if desired.

- Collect the coagulated curd by filtering the mass through a towel or a filter.
- If you press the curd you will get a hard cheese; if the curd is pressed lightly or not at all, the cheese is consumed fresh.
- Add some salt if you like (if you did not do so before)
- Store the cheese in the fridge.

Whey as Animal Feed

If whey is discarded it can cause serious problems as a dairy effluent. Therefore it is far better to use the whey as an animal feed. It has a good feeding value because of the presence of some of the whey proteins. It can be used for feeding pigs or young animals like calves or lambs.

Even adult cows or fattening cows can be fed with whey. It is important that the whey used for animal feed is acid (completely acidified) at the moment of feeding, otherwise the animals can get intestinal problems because of the lactose content of the whey.

UTENSILS FOR MAKING CHEESE

When making cheese, you will not always need all the utensils described below. Decide beforehand which you will use, and make sure that they are clean and rinsed with clean water. It is preferable to use utensils made of stainless steel or glass.

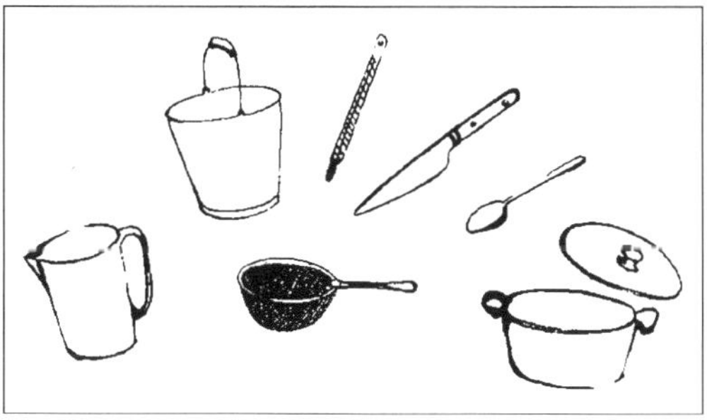

Fig. Utensils for Cheese Making

The utensils used during cheese making are:

- A thermometer with a range of 20–100°C
- A measuring cup
- A bucket for coagulation of the milk
- Cheese moulds, which can be made in various ways
- Cheesecloth, its size depends on the size of the mould
- Cutlery and other tools such as:
  - Spoons to measure rennet and/or acids
  - A knife to cut curdled milk
  - A skimmer to scoop curd out of whey, or a colander.

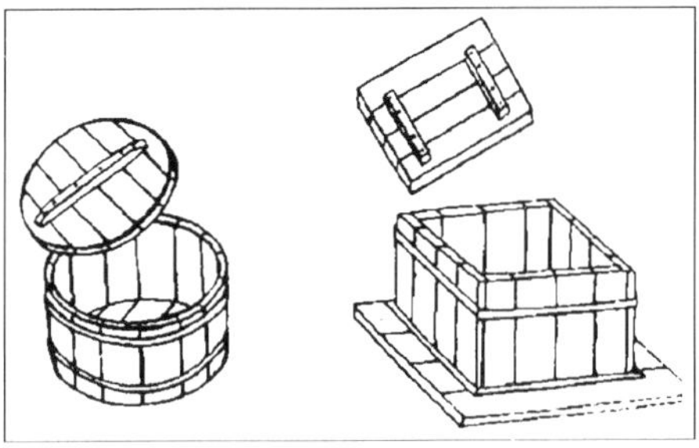

Fig. Cheese Moulds

Cheese moulds can be made from various materials *e.g.* wood, plastic or metal. Metal materials should be non-rusting. Do not use plastic tubing used in construction as it can release poisonous substances. Cut the moulds lengthwise and drill holes from the inside out. Plastic moulds for the production of soft cheese are also commercially available. Wooden cheese moulds can be used as well. Cheese that will have a closed rind after pressing is never placed directly into a mould. Instead, a clean cloth of linen or cotton is first placed in the mould and it is then filled with curd. The cheesecloth is then folded closed.

To make hard or semi-hard cheese you will also need the following:

- A curd knife to cut the renneted milk
- A cheese press

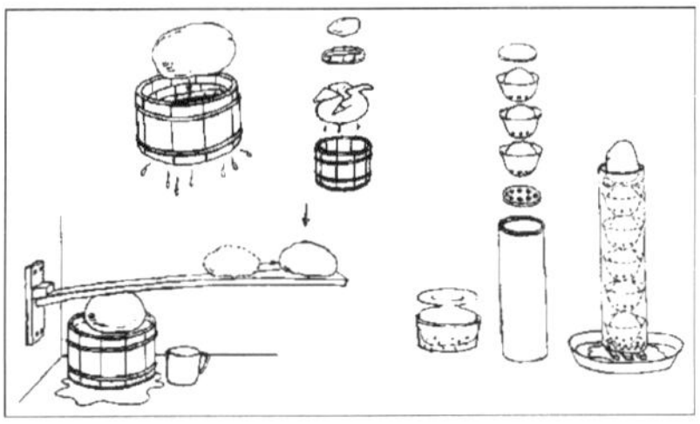

Fig. Some Examples of Cheese Presses

Buy these second-hand if possible or make your own simple press, paying special attention to the following:

- The materials used must not be poisonous.
- It must be possible to easily clean the materials.
- It must be possible to build up sufficient pressure (0.1 to 0.4 kg/cm2 or 2-5 x the weight of the cheese).

THE QUALITY OF MILK USED TO MAKE CHEESE

Cheese making begins with hygienic milking. The quality of the milk used greatly influences the smell, taste and keeping qualities of cheese. The composition of milk can vary greatly. Just before a cow is set dry–the period in which a cow is not milked–and just after calving, the cow milk has a different taste and composition. Milk from a cow with an udder infection is not suitable for human consumption, and therefore cannot be used for making cheese.

Apart from the usual measures taken to ensure good hygiene during milking, attention must be paid to the following points:

- Clean the equipment very well.
- Disinfectants and cleaning agents must never be left in or on equipment. Disinfectants inhibit the growth of the bacteria of the starter or souring agent.
- The room in which cheese is prepared must be kept properly clean.
- The milk used for cheese making should preferably be pasteurised (15 seconds at 72°C or 30 minutes at 63°C). More intense heating is not desirable because it reduces the curdling of milk. More rennet or calcium chloride (CaCl2) is then needed.
- For the above reasons, milk powder which is strongly heated during its production (high-heat milk powder) is less suitable for cheese processing. Only low-heat milk powder can be used.
- Sour milk neutralised with sodium bicarbonate will not curdle well.

COAGULATION OF THE MILK

The principle of cheese processing is based on the coagulation of the protein in milk, during which about 90% of the milk fat is encapsulated. The coagulated mass is called curd, the remaining liquid is called whey. Curd consists mainly of milk proteins (casein) and milk fat; while whey mainly contains water, milk sugar (lactose), protein (serum proteins) and B-vitamins.

There are two basic ways to cause milk to coagulate:

1. Using an acid
2. Using a rennet (enzyme coagulation)

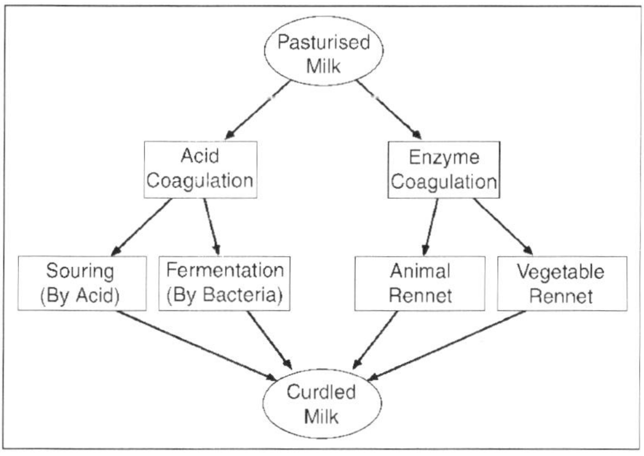

Fig. Four Methods of Curdling Milk

Acid Coagulation

This is mainly used to make fresh cheese. The acid can come from bacteria from the starter culture or from an acid added to the milk. When using a culture, pasteurised milk is inoculated. Inoculating agents can be a specific cheese starter culture, whey or sour milk. Curdling time depends on the amount of inoculating agent added (0.1-5% of the milk used to make cheese), temperature (20-35°C) and the starter culture used. It takes 2-16 hours. When the curd is firm, the curdling is finished.

When using an added acid, you can use pure vinegar acid, lactic acid, citric acid or any other harmless organic acid. Sometimes a natural acid such as lemon juice is added.

The acid can be added drop by drop to warm milk (about 80-90°C, just after boiling). Milk curdles much quicker at a high temperature. The curd is collected by straining the curdled milk through a coarse cloth. Sometimes the collected curd is pressed a little bit, and sometimes no pressure is applied. These products are consumed fresh.

Rennet

Rennet used in cheese processing can be of animal, vegetable or microbial origin.

The rennet in cheese has two functions:

- Rennet causes the milk to coagulate.
- During ripening, rennet causes milk protein to break down, giving the typical cheese taste.

Rennet is available in liquid form and as a dried powder. As it is an enzyme and therefore a biological product, the strength of the liquid product decreases during storage. Therefore, it is preferable to use the dried form. The concentration of the rennet is written on the package.

The amount of rennet to be added depends on the strength of the rennet and on the kind of cheese you are going to make. When making soft cheese, sometimes no rennet is needed; if rennet is added, only a little is needed (0.1 ml of rennet to 10 litres of milk). For hard or semi-hard cheeses, about 1.5 ml of rennet is added to 10 litres of milk; assuming the rennet concentration is 1:10,000. A starter is almost always added before or while adding rennet.

A number of factors influence coagulation:

- Amount of rennet or acid added.
- *Curdling Temperature*: When using rennet a small rise in temperature (for example from 30°C to 33°C) can significantly reduce coagulation time.
- Intensity of pasteurisation. Curdling is reduced when milk has been strongly heated. This effect can be neutralised by adding a small amount of CaCl2; for example 7 grams of CaCl2 per 100 litres of milk.
- *Fat Percentage of the Milk*: A high fat percentage means more fat must be encapsulated and curdling occurs a little bit slower.

Preparation of Rennet

Rennet from animal origin is prepared from the stomach of suckling calves or lambs. It is obtained by extraction of the stomachs in a salt solution with a preservative. *A traditional method formerly used in The Netherlands can be applied in the tropics:*

- Take 9 litres of boiled water.
- Add 26 carved calf stomachs, 500 gram sodium chloride (salt) and 200 gram boric acid.
- Store these for at least 10 days and stir once per day.
- After ten days, remove the stomachs from the solution, add another 500 gram salt to the solution and store it for some days more.
- Then, pour it into clean bottles, close these and store the bottles in a cool and dark place. Your rennet is ready.

You have to try out the strength of the rennet. Also, the keeping quality of these rennets is usually poor. Sometimes the stomachs of the young animals are kept after cleaning, salting and drying. The dried stomachs are easy to transport and are used by some nomadic tribes. In cheese making a piece of dried stomach is then added to the milk.

CHEESE RECIPES

There are many variations in cheese making. To make a good product, it may be necessary to adjust the recipe. Therefore you should have a good method of recording exactly how the cheese is made.

Your records can include the following aspects:

- Date and surrounding temperature
- Quality of the milk and pasteurisation temperature
- Aamount of milk
- Ingredients, amount of starter culture, acid or rennet added
- Temperature at which starter culture, acid or rennet was added
- Coagulation time
- Temperatures during the cheese making process
- Temperature at the end of coagulation
- Pressure applied, length of time applied, etc.
- Salting time
- Storage time and conditions during storage

The following recipes should be seen as guidelines for making cheese. Instead of cow milk, you can often use goat, sheep or buffalo milk. Start by making rather simple products like yoghurt, fresh cheese, etc.

Fresh Cheese

Fresh or unripened cheese has a high moisture content of about 75%; it can be consumed directly after preparation. It is made by removing the whey from soured, skimmed milk. The milk is usually coagulated by souring. Sometimes a small amount of rennet is added; this is done to facilitate the draining of the whey. However, addition of rennet is not essential.

A well-known kind of fresh cheese is curd, known in various countries by the following names: *Frischkäse*, *fromage frais*, quarg, and baker's cheese. The differences between these cheeses are their fat levels. You can make curd by souring fresh milk, cream or skimmed milk into sour milk, sour cream or sour skimmed milk and then draining the thick sour milk in a bag or cloth. Sometimes bags are placed on top of each other to increase the removal of whey.

After draining, the curd has a crumbly structure. By stirring or using a blender you can make the product smooth again. There are various kinds of curd cheese in which moisture, fat and salt content and size of curd particles vary. Fresh cheese has a fresh sour taste, especially when it is prepared from skimmed milk. By adding cream, the taste will become milder and richer. Curd can be kept for only a short time and must be stored at cool temperatures. During storage, further whey separation can occur; this is the result of further souring of the product.

Curd from Whole Milk

Pasteurise the milk 30 minutes at 63°C and then cool it to 20°C. Add per 10 litres of milk ¼-½ litre of starter culture or fresh sour milk or buttermilk; yoghurt can be added if desired. Add 2 drops of rennet-if available. It is advisable to dilute this small amount of rennet with several ml of water to improve its distribution through the milk. After stirring well, leave the inoculated milk to stand for 24 hours at 18–20°C. During those 24 hours, souring and coagulation will take place; the milk becomes a rather firm mass. This thick mass is subsequently poured into a cotton or linen cloth or a bag so that the whey can drain through the cloth. The cloth is placed into a large colander or cheese mould beforehand so that the whey can thoroughly drain. After 24 hours, sufficient whey will have leaked out and the remaining curd can then be mixed, for example with a spoon or blender, until it becomes a homogenous mass. The curd can now be consumed. When kept in a refrigerator, curd can be stored for 1-2 weeks.

Bag Cheese

You will need a sieve, small basket or cheese mould and a cheesecloth or tea towel. Let 10 litres of buttermilk with low fat content drain through a cheesecloth until 1.5 litres of bag cheese, or curd, remain. Place a cheesecloth in a sieve, small basket or cheese mould and press the curd firmly into it. Let the curd drain for several hours, then turn it. About 1.5 kg of bag cheese has then been made, which contains no salt and hardly any fat. It can be kept for only a rather short time: 1 to 2 weeks. Keep it cool, preferably in a refrigerator.

Fig. Bag Cheese

Krut

For making krut one needs buttermilk, cheesecloth and salt. This is a way to make cheese from any leftover milk. Add some sour milk or buttermilk to the leftover milk and mix. Boil the mixture until the milk curdles. Separate the curd from the whey by pouring it through a cheesecloth. The curd is then kneaded with 2-4% salt and dried in the sun.

Rasagollas

You will need raw milk, a fire, pan, spoon, lemon juice or sour whey, cheese-cloth, bowl, knife and concentrated sugar water. Rasagollas is a sweet dairy product originally from India. Traditionally it was formed into sweet curd balls, but because of the way it is prepared here, the curd can only be cut into cubes and not formed into balls. The milk is boiled with lemon juice (10 tablespoons or 150 ml per 10 litres of milk) or sour whey (1.5 litres per 10 litres of milk) while stirring continuously. Sour whey can be obtained from drained sour milk after the production of curd.

The curd is separated from the whey by pouring the mixture into a cheesecloth, which has been placed over a bowl. The slightly elastic curd is cut into cubes with sides of about 2.5 cm. These are then boiled for about an hour in a sugar solution of 600 g per litre of water. 1 litre of sugar solution is needed for each kilogram of curd. The cubes can be stored quite some time and are very sweet.

Feta

You will need sheep, goat or cow milk, starter culture or fresh sour milk, rennet, cooking-salt, knife, cheese moulds, cheesecloths and cans or a plastic container to store the cheese. Feta is a sharp, salty cheese originally from Greece that is made of sheep and goat milk. You can also use a mixture of sheep and cow milk, but then the cheese will not have its typical white colour.

Feta is kept in a solution of whey and brine. Heat 10 litres of pasteurised milk mixed with 200 ml of a starter culture, sour milk or buttermilk to 30°C. After 1-2 hours, rennet is added; use 2 ml of rennet per 10 litres of milk. After allowing it to coagulate for about one hour, the curd is cut into cubes of about 2.5 cm, after which it is carefully stirred for another 20 minutes. The curd must then be carefully transferred to the tubs covered with cheesecloth.

This can be done either by scooping the curd directly out of the whey into the moulds lined with cheesecloth, or by letting the curd settle, pouring off the whey and only then putting the curd in the cheese moulds. After a few hours the cheese must be turned. The curd mass is carefully removed from the cheesecloth and replaced upside down. After one day, the lumps of curd are cut into cubes of about 10 cm.

Salting can be done by sprinkling the blocks several times with salt or by placing the blocks in brine for 24 hours. If the cheese is kept for several days at about 18°C, it must be turned regularly and washed with cold water at the end of the storage period. The cheese can be kept for some time by piling blocks of cheese closely on top of each other and covering them with brine. The cheese should have a smooth and soft consistency.

Queso Blanco

The so-called *queso blanco* is manufactured in many ways in Latin America. Typical for this cheese is that salt is added directly to the whey/curd mixture. This has the advantage that slightly soured milk can be used to make *queso blanco*.

A much-used method is the following:

- Take raw (unheated) soured milk of 32°C, or take 10 litres of pasteurised milk and add 50 ml of sour milk, buttermilk or starter culture. Add 1.5 ml of rennet. After 45 minutes the curd is cut and stirred. Leave the whey/curd mixture to settle for another 30 minutes at 30-36°C. Pour off the whey and compress the curd to remove more whey. Mix 30-50 grams of salt through the curd. The salt can also be dissolved in water before being added. 100 grams of salt dissolved in 50 ml of water must then be added for each 10 litres of milk.

Transfer the salted curd into cheese moulds and press the cheese. During the first hours, turn the cheeses occasionally. Press them until the following day. To improve rind formation, pour whey heated at 50°C over the cheese while being pressed (after one hour). Cheese made with rennet can be kept for 2 months at 10-15°C. As a variation to this method, you can add acid instead of a starter culture and use no rennet. Take raw milk, which may already be a little bit sour. Heat it to almost boiling. Acidify it with 300 ml vinegar per 10 litres of milk until a precipitate is formed. The acid can be partly neutralised with the addition of some sodium bicarbonate (double soda). Pour off the whey.

Fresh Goat Cheese (in Oil)

You will need pasteurised goat milk, a heat source, a pan with lid, a thermometer, sour milk (or buttermilk or starter culture), rennet, a spoon, an insulated box or blanket or newspapers, cheesecloth, salt, cheese moulds, a cool storage place, greaseproof paper (if available), a large pot, herbs and olive oil. Bring the pasteurised goat milk to a temperature of 20°C. Per 10 litres of milk, add 0.5 litre of fresh starter or sour milk or buttermilk.

Take 20 drops of rennet per litre of milk, dilute this in a little water and stir this through the milk. Place the pan in the insulated box or in the blanket or the newspapers to prevent it from cooling down. Check the next day to see if the milk has curdled sufficiently; a little whey on top is acceptable. Cut the curd into cubes the size of a matchbox. After 2 hours transfer the mass into a cheesecloth and let it drain for 12 hours at room temperature.

Mix the dry curd with some salt and put the cubes into cheese moulds. Press the curd into the moulds so that no air holes remain and place the moulds in a cool (15°C) place. The next day remove the cheese carefully from the cheesecloth and replace it upside down in the cheesecloth and mould. Leave the cheese for a further 24 hours in a cool place.

Remove the cheeses from cloth and mould and turn them daily. Should they become too dry, wrap them tightly, for example in greaseproof paper. If you wish to store the fresh goat cheese cubes for several weeks, place them in a large (glass) jar, which can be closed. Sprinkle a mixture of different herbs over them, such as rosemary, basil, thyme, bruised juniper berries, a finely chopped piece of garlic, several pepper kernels and a chopped up hot pepper.

Other herbs can also be used. Pour olive oil over the cheeses until they are just covered and add a few twigs of dill or rosemary to the oil. Close the jar and put it away in a cool and dark place. Before using the cheeses, let them drain well. Use the leftover oil for salad dressing.

Fresh Goat Cheese (Salted)

You will need pasteurised goat milk, a heat source, pan with lid, thermometer, sour milk (or buttermilk or starter culture), acid, spoon, rennet, knife, colander or cheesecloth, cheese moulds, pressing equipment and salt. Heat the pasteurised milk in a pan to 30°C, stirring continuously. For each 10 litres of milk, add 0.2-0.5 litres of sour milk or buttermilk or 0.2 litres of a starter culture and 30 drops of rennet (diluted in water).

After 45 minutes, cut the coagulated curd with a knife; after another 10 minutes, the curd will be the size of a marble. The top layer of whey can be poured off after leaving the mixture to stand a bit. Leave the curd in the carefully closed pan with the rest of the whey for a further 30-45 minutes. Then transfer the curd into cheese moulds in which cheesecloth has been placed. Press the curd; one hour is sufficient. The cheeses must then be pickled in brine. For a cheese of 500 grams, 10 hours of pickling is long enough.

Fresh Sheep Cheese

You will need pasteurised sheep milk, a heat source, a pan, a thermometer, a starter culture or fresh sour milk or buttermilk, rennet, a spoon, a knife, a cheesecloth, some salt (if desired) and cheese moulds. Sheep milk contains more fat and protein than cow milk and therefore only 4-4.5 litres of sheep milk are needed to make 1 kg of cheese. Because of the higher content of dry matter in the milk, sheep milk coagulates firmer than cow milk.

Heat the pasteurised milk to about 30°C, stirring continuously. Add 300 ml of a starter culture and 40 drops of rennet to each 10 litres of milk. After 45 minutes the milk will have coagulated sufficiently to be cut. Cut the curdled, thick mass carefully; continue till the curd particles have the size of a pea. Bring the curd into a cheesecloth. If desired mix in some salt, and let the cloth hang until the cheese has the desired firmness.

You can speed up the draining of the whey by opening the cheesecloth after several hours, scraping the thick part from the cloth, and mixing it in with the rest of the curd. You may want to make a less sour sheep cheese, which is a little bit firmer and with some lower moisture content. For this, you must leave the curd for a longer time: about 15-30 minutes in the cheese vat where the curd treatment is done. After a short period of rest, pour off part of the whey, stir the curd carefully again and fill the cheese mould using a cheesecloth. Add some salt to the curd, if desired. Press the curd: about 4 hours of light pressing is sufficient. Store in a cool place. Sheep cheese can be kept for at least one week in the refrigerator.

Matured Sheep Cheese

You will need pasteurised sheep milk, a heat source, a pan, a thermometer, sour milk (buttermilk or starter culture), rennet, a spoon, a knife, colander, a cheesecloth, cheese moulds and pressing equipment, salt, a tightly sealed pickling tub, coarse cheesecloth and a cool storage place. 10 litres of sheep milk yields about 2 kg of ripened cheese. After adding 60 ml of starter culture or sour milk (or buttermilk), leave the milk to stand at 30°C for 30-45 minutes.

Only then add 60 drops of rennet, diluted with a little water, per 10 litres of milk, and stir it thoroughly through the milk. After allowing it to curdle for 1 hour, cut the coagulated milk for about 15 minutes until it is divided into particles of about 1-2 cm. Then stir it for 10 minutes, pour half of the whey off and heat the curd to 35°C by adding hot water with a temperature of 80-100°C.

Stir the mixture again for about 15 minutes, after which time the curd must be left to stand for 30 minutes in the pan. Keep the pan warm as well as possible but never place it on a fire. Remove the whey that has separated, and pour the curd with the remaining whey into a colander. After the first draining, the curd must be transferred by hand into cheese moulds covered with cheesecloth.

FLAVOUR LEXICONS FOR DAIRY PRODUCTS

M.A. Drake and G.V. Civille have reviewed lexicon history, methods and applications. A flavour lexicon is a set of word descriptors that describe a product's flavour. While the panel generates its own list to describe the product array under study, a lexicon provides a source of possible terms with references and definitions for clarification.

Development of a representative flavour lexicon requires several steps, including appropriate product frame-of-reference collection, language generation and designation of definitions and references, before a final descriptor list can be determined.

Once developed, flavour lexicons can be used to record and define product flavour, compare products and determine storage stability, as well as to study correlations of sensory data with consumer liking/acceptability and chemical flavour data. Good flavour lexicons should be both discriminating and descriptive.

The language should be developed from a broad representative sample set that exhibits all the potential variability within the product. For example, Drake collected 220 samples of Cheddar cheese varying in age, milk heat treatment and geographical origin to identify a descriptive language for Cheddar cheese.

The sample set was screened to 70 cheeses prior to language generation. In creating a lexicon, the panel will frequently review the list, merging like terms, eliminating redundancies and organizing the list so that the attributes appear in most products being tested. It is important that multiple terms are not used to describe the same flavour; conversely, it is also important that one term doesn't represent or overlap with several other flavours.

As an example of this type of lexicon problem, Drake reported that use of the term "aged" in a Cheddar cheese flavour lexicon was in fact a meta term that comprised three flavours and one basic taste.

An optimized lexicon can relate consumer acceptance/rejection and instrumental or physical measurements. The use of chemical components, particularly those isolated from the product under study, can make a lexicon clearer and more grounded, establishing a link to the formulation and/or production of that product.

Creating this type of link can be time-consuming and challenging. However, even without such chemical references, a lexicon can be discriminating and precise. Several diff e rent flavour lexicons have been developed to study cheese aroma and flavour development, the effects of fat reduction and the effects of different starter or adjunct bacteria. Muir described nine aroma terms for characterization of aroma profiles of hard and semi hard cheeses.

For studying flavour development in Cheddar cheese during maturation, Piggot and Mowat determined 23 descriptive flavour terms, and Roberts and Vickers developed a flavour lexicon. Muir and Drake used descriptive sensory panels to determine the effect of starter culture and adjunct cultures on Cheddar cheese flavour. Banks used descriptive analysis to determine sensory properties of low fat Cheddar cheese.

APPLICATION OF SENSORY ANALYSIS TO DAIRY PRODUCTS

Following are specific examples of how QDA and/or other types of sensory analysis techniques have been applied to dairy research studies funded by Dairy Management Inc. ™ (DMI).

OPTIMIZATION OF CHEDDAR CHEESE TASTE IN MODEL CHEESE SYSTEMS

Cheddar cheese, the most popular natural cheese in the United States, has a very complex flavour system. While much information has accumulated during the past century, the industry is still seeking to fully understand Cheddar cheese flavour and has not been able to replicate it in model systems.

The nonvolatile sensory attributes of Cheddar cheese are important for providing the character of Cheddar cheese.

While much work has been published on volatile components of Cheddar, far less is known on how nonvolatiles impact Cheddar flavour. Yang, a Kraft Foods researcher, and Vickers, at the Minnesota-South Dakota Dairy Foods Research Center, used sensory analysis to better understand the importance of nonvolatile compounds to Cheddar flavour.

A descriptive panel was trained to evaluate real and model cheese for a variety of taste attributes and for Cheddarlike taste. Sodium chloride, lactic acid, citric acid and monosodium glutamate were added to the model systems using mixture designs and response surface methodology to determine optimum levels of these components.

The three model systems investigated were:

1. A dairy model system (containing milk isolate, anhydrous milk fat, water, annatto colour and chymosin);
2. A nondairy model system (containing gelatin, gum acacia, modified starch, sunflower oil, water and annato colour); and
3. A mozzarella base.

While the mozzarella base did present tastes, it was used because the other two model systems were too unlike Cheddar cheese (or any cheese) texture. Less sodium chloride and fewer acids were required to simulate the taste of mild Cheddar compared with aged Cheddar. None of the model systems mimicked the texture of real Cheddar. The researchers were able to match approximately, but not exactly, the taste of aged Cheddar using a mozzarella base.

Panelists generally rated the optimized taste in the dairy model system as more Cheddar-like than the optimized tastes in the nondairy model. Two methods were used to measure how close a sample was to the Cheddar concept. One was by measuring the similarity of the sample to either mild Cheddar cheese taste or aged Cheddar cheese taste on an unstructured scale.

The left end of the line was marked with "not at all like Cheddar taste" and the right end was marked with "exactly like mild Cheddar taste" or "exactly like aged Cheddar taste." The other method was by concept matching using an R-index methodology. For the mild group, the panel evaluated whether the samples were "MC" (mild Cheddar taste and sure), "MC?" (mild Cheddar taste but not sure), "N?" (no Cheddar taste but not sure) or "N" (no Cheddar taste and sure). For the aged group, "MC" and "MC?" were changed to "AC" (aged Cheddar taste and sure) and "AC?" (aged Cheddar taste but not sure).

A model system for studying Cheddar taste should be as bland-tasting as possible, and also have a texture and composition similar to that of real Cheddar cheese. The characteristic flavour of a food depends not only on the flavour compounds present and their levels but also the rate and extent to which they are released in real time, which in turn are affected by the amounts of proteins, fat and other matrix components of the sample.

By using a trained descriptive analysis panel, Yang and Vickers were able to evaluate the flavour impact of several nonvolatile Cheddar cheese components (*i.e.*, salt, lactic acid, citric acid and monosodium glutamate) in model systems that attempted to mimic real Cheddar cheese. They achieved the most Cheddar-like taste with the mozzarella cheese base, and panelists found the optimal concentration of salts and acids in the model to be nearly indistinguishable from real Cheddar cheese.

CHEDDAR CHEESE AND POWDERED MILK LEXICONS

M. A. Drake, at the Southeast Dairy Foods Research Center, developed and validated a descriptive language for Cheddar cheese flavour. For the project, 240 representative cheese samples were collected. Fifteen individuals from industry, academia and government participated in roundtable discussions to generate descriptive flavour terms. A highly trained descriptive panel (n–11) refined the terms and identified references.

Identification of chemical references was conducted with the assistance of K. Cadwallader at the University of Illinois. Instrumental analyses (gas chomatography/mass spectrometry, or GC/MS) were conducted to identify many flavour compounds that were responsible for specific flavours and off-flavours in Cheddar cheese. Twenty-four Cheddar cheeses were then presented to the panel to validate the proposed lexicon.

The panel differentiated the 24 Cheddar cheeses as determined by univariate and multivariate analysis of variance. Twenty-seven terms were identified to describe Cheddar flavour. Seventeen descriptive terms were observed in most Cheddar cheeses. Drake's standard sensory language for Cheddar cheese today is facilitating training and communication among different research groups. The Cheddar cheese lexicon is helping cheese-makers and cheese users accurately and consistently characterize the flavour of their cheese products and improve quality issues by measuring and controlling the presence of compounds that have been associated with flavour defects.

Following development of the Cheddar cheese lexicon, Drake developed a similar language to help characterize another food industry staple: dried dairy ingredients, including whey proteins and nonfat dry milk. Global production of nonfat dry milk tops 3.3 million tons and whey protein demand still outstrips production, which increases annually.

A sensory lexicon describing the flavour of these ingredients helps dairy processors maximize the quality of these ingredients and allows food technologists to identify the exact attributes or flavour notes these ingredients contribute to formulations. Drake said she was surprised by the number of descriptive terms that the panel uncovered for application to the dried dairy ingredients lexicon. The panel discovered 21 flavour terms that could be applied to milk powders. Examples included cooked/milky flavour, cake mix or vanillin, sweet and sour, earth and cereal.

Each of these flavours was linked to a key aroma compound, many of which were identified by Drake and Cadwallader with GC/MS. For example, lactones tend to lend a sweet, coconut like flavour, while various free fatty acids can simulate a waxy flavour. Many different factors contribute to flavour variability. The source of the powder, processing/packaging methods and materials, as well as storage time and conditions, are just a few.

The dried dairy ingredient lexicon, linking responsible chemical factors and causal agents, provides common ground for processors and ingredient suppliers to discuss ingredient characteristics. Figure shows how descriptive analysis results based on Drake's dried dairy ingredient lexicon can be analysed by PCA. This two-dimensional PCA plot shows the attribute variability among 27 low-heat skim milk powders less than three months old. Rehydrated milk powders are represented by numbers. PC1 = principal component 1; PC2 = principal component 2.

PROCESSING OF OTHER CREAMS

WHIPPED CREAM

Whipped cream contains added sugar and stabilisers. The stabilised cream is then pasteurised and held at 5°C for 24 hours. Compressed air or nitrogen is then introduced into the mix. This provides an excellent aerobic medium for microbial growth, thereby increasing the chances of spoilage in comparison with liquid cream.

FROZEN CREAM

Frozen cream is pasteurised at ≥ 75°C for 15 seconds. It is then quickly cooled to 1°C before being frozen in containers, as sheets or pellets, or by direct contact with liquid nitrogen. It is then stored at-18 to-26°C.

CLOTTED CREAMS

Clotted creams are traditionally made by putting milk in a pan 30 cm in diameter and 20 cm deep where it is held for 12 hours to allow the cream to rise. The pan is then put on a steamer until a layer of solidified cream is formed around the edge.

Modern methods involve:

- Heating double cream over a layer of skimmed or whole milk, in a large, shallow jacketed tray until a crust is formed or
- Heating a thin layer of high-fat cream (54–59% milk fat) at 77–85°C, to form a crust. The more severe heat treatments result in aerobic spore-formers being the predominant microflora.

However, slow cooling and poor hygiene are more likely to lead to the growth of spoilage moulds, coliforms and other contaminants.

CREAM-BASED DESSERTS

Cream-based desserts typically undergo heat treatments above pasteurisation, in order to allow cooking of other ingredients such as starch.

SPOILAGE

The spoilage of cream is generally similar to that described for liquid milk products. However, because of the difference in purchasing patterns, cream is often required to have a longer shelf life than milk (up to 14 days for pasteurised cream), and containers may be opened and then used by the consumer over several days. The keys to obtaining sufficient shelf life are the microbiological quality of the raw milk, good hygiene in processing, and effective temperature control during distribution and storage.

Cream usually receives more severe heat processes than milk, and the post heat treatment microbial population therefore consists almost entirely of relatively heat-resistant species. Aerobic spore-forming bacteria survive pasteurisation, and psychrotrophic strains of *Bacillus cereus* may cause 'sweet curdling' and 'bitty cream'. Other, more heat-resistant species, such as *Bacillus licheniformis*, *Bacillus coagulans*, and *Bacillus subtilis*, may survive sterilisation and even UHT processes, and may cause bitterness and thinning in sterilised creams.

Bacillus pumilus and *Bacillus sporothermophilus* are now recognised as potential contaminants in cream, primarily carried over from raw milk. Under UHT conditions, *B. sporothermophilus* has D-values of 3.4-7.9 sec and z-values of 13.1-14.2. Heat-resistant lipases produced by psychrotrophic bacteria growing in the raw milk may also survive high-temperature processing and cause spoilage in UHT cream. The keeping quality of cream is greatly affected by the introduction of postprocess contamination. Psychrotrophic bacteria such as pseudomonads may contaminate pasteurised cream during processing and are important spoilage organisms. The high fat content of cream means that lipolytic species, such as *Pseudomonas fluorescens* and *Pseudomonas fragi*, are a particular problem. A study of pasteurised double cream showed that pseudomonads were the predominant spoilage organisms. Psychrotrophic members of the Enterobacteriaceae are also sometimes involved.

Yeasts and moulds are rarely implicated in the spoilage of cream. Few yeasts are able to ferment lactose, but species such as *Candida lipolyticum* and *Geotrichum candidum* may occasionally spoil bakers' whipping cream where sucrose has been added. If, however, other organisms hydrolyse lactose, then the yeast can grow rapidly to produce yeasty or fruity flavours and gas; *Torula cremoris*, *Candida pseudotropicalis* and *Torulopsis sphaerica* have been implicated with such defects. Where cream is stored at very low temperatures (0-1°C) to prolong the shelf life, mould growth, usually *Penicillium* spp. may develop on the cream surface.

Defective cans or leaking seams could cause spoilage of cream due to entry of bacteria from cooling water or other sources, *e.g.* a waterborne organism, for example *Proteus*, can cause bitterness and thinning, coliforms can produce gas, and lactococci could result in acid curdling.

In the case of cream-based desserts, thermoduric organisms are most likely to be an issue due the more aggressive heat treatments that are used. In addition, the added sugar increases the range of contaminants that could grow in the product. Fruit conserves, if added, will lower the pH of the product thus favouring the growth of yeasts and moulds. With multi-component desserts, both individual components, and blends obtained from their mixing could be responsible for microbial spoilage.

PATHOGENS: GROWTH AND SURVIVAL

In practice, to overcome the protective effect of the higher fat content, cream usually receives a more severe heat treatment than milk. This means that pathogens present in the raw cream are more likely to be destroyed. Unpasteurised cream carries a high risk from the presence of foodborne pathogens, as does raw milk, but the recent safety record of pasteurised cream is good. Although foodpoisoning outbreaks have been associated with cream, they are often linked to products filled with, or prepared with cream. In these cases, it is probable that poor hygiene during manufacture, and temperature abuse during storage have been important contributory factors.

SALMONELLA SPP.

Salmonellae will not survive the heat treatment applied to cream, and therefore their presence is likely to be due to post-pasteurisation contamination. The cells are likely to survive for extended periods in contaminated cream, but growth is not possible unless significant temperature abuse occurs. Storage at temperatures below 5°C will prevent multiplication. Most of the relatively recent outbreaks recorded have been associated with foods prepared with cream. For example, in 1986 an outbreak of *Salmonella typhimurium* DT40 infection affecting 24 people in the UK was linked to consumption of cream-filled profiteroles.

A much larger outbreak occurred in Navarra in Spain in 1991, and was reported to have affected approximately 1,000 people. The causative organism was *Salmonella enteritidis*, and the outbreak was associated with the consumption of contaminated confectionery custard and whipped cream.

In 1992, an outbreak of *S. enteritidis* PT4 infection in Wales was associated with fresh cream cakes, and was found to be a result of contamination of the factory environment by the organism, and inadequate cleaning of the nozzles used to pipe cream into the cakes. More recently, in 1998, an outbreak of *S. typhimurium* DT104 infection affected 86 people in Lancashire. The outbreak was linked to inadequately pasteurised milk from a local dairy, but cream from the same dairy was also recalled.

LISTERIA MONOCYTOGENES

There has been some concern that *L. monocytogenes* might be able to survive cream pasteurisation processes and then grow during chilled storage. However, *L. monocytogenes* strain Scott A recorded a D-value of 6 seconds at 68.9 °C in raw cream with a fat content of 38%, indicating that pasteurisation is likely to be effective. The D-value increases to 7.8 sec in inoculated 'sterile' cream. Z-values were calculated as 6.8°C and 7.1°C, respectively.

A later study, using two strains of *L. monocytogenes* suspended in different dairy products, including half and double cream, showed that, although heat resistance did vary, minimum pasteurisation processes would be adequate to eliminate the organism in all products. An investigation into the fate of several strains of *L. monocytogenes* in whipping cream at various storage temperatures recorded generation times of 29-46 hours at 4°C.

Populations of approximately 107 cells/ml were reached after incubation for 30 days, and, at 8°C, hazardous levels were reached in only 8 days. This indicates that post-pasteurisation contamination of cream could be a potentially serious problem. The same post-process hygiene precautions should be applied for cream as for other high-risk chilled products. Despite this, although *L. monocytogenes* infection is reported to have been linked to cream on epidemiological evidence, such cases have not been confirmed by microbiological investigation.

YERSINIA SPP.

Yersinia enterocolitica is a common contaminant of raw milk, although the majority of the strains isolated are not pathogenic to humans. The organism is heat-sensitive and does not survive pasteurisation, but is capable of psychrotrophic growth. Therefore, it is a potential hazard in cream if introduced as a post-pasteurisation contaminant. A survey of dairy products in Australia recorded an isolation of *Y. enterocolitica* from pasteurised cream, but the organism was not detected in cream sampled in the UK over the course of a 3-year survey to determine its incidence in foods. There have been no reported outbreaks of *Y. enterocolitica* infection associated with cream.

STAPHYLOCOCCUS AUREUS

Although *Staph. aureus* can often be isolated from raw milk, and is a common cause of mastitis in cows, it does not survive pasteurisation, and cases of staphylococcal food poisoning from pasteurised dairy products are now uncommon. It may be introduced into cream as a post-process contaminant, particularly from infected food handlers. However, it is incapable of growth below about 7 °C, and high numbers will only develop following significant temperature abuse. An investigation of growth and enterotoxin.

A production by *Staph. aureus* in whey cream showed that growth was limited and that enterotoxin was not produced at detectable levels. Despite this, between 1951 and 1970, six outbreaks of staphylococcal poisoning associated with cream were recorded in England and Wales. There have been few recent reports of outbreaks, following significant improvements in hygiene and temperature control. As with *Salmonella*, products prepared or filled with cream are now more likely to be implicated as vehicles of staphylococcal poisoning than cream itself, usually as a result of poor hygiene during handling and temperature abuse.

BACILLUS CEREUS

B. cereus is common in milk, and its endospores are able to survive pasteurisation. Some strains are also psychrotrophic, and capable of growth in refrigerated dairy products. Nevertheless, there are very few reports of *B. cereus* food poisoning associated with dairy products. There have been a small number of outbreaks associated with the consumption of pasteurised cream. In 1975 cream found to contain 5xl06 cfu *B. cereus* caused illness in several people. In 1989, two members of the same family became ill after consuming fresh single cream that was later found to contain *B. cereus* at levels of 3x107/g.

VEROTOXIGENIC ESCHERICHIA COLI (VTEC)

VTEC, particularly *Escherichia coli* O157, have been found in raw milk and have caused serious outbreaks of infection associated with consumption of raw or inadequately pasteurised dairy products. An outbreak of *E. coli* O157 infection was recorded in the UK in 1998, associated with consumption of raw cream from a small farm dairy.

Seven cases were recorded, with four requiring admission to hospital. These organisms are destroyed by properly applied pasteurisation, but if any opportunities for cross-contamination between raw and pasteurised cream exist, recontamination could potentially occur. It is likely that *E. coli* O157 could survive for prolonged periods in cream, but growth in the absence of temperature abuse is improbable. In view of the potentially serious nature of infections caused by VTEC, and the low infective dose, it is important to ensure that such cross-contamination does not occur, since growth may not be required to cause infection.

VIRUSES

Viral hepatitis is the most likely viral infection to be associated with dairy products. In 1975 in Scotland, an outbreak of hepatitis A infection occurred associated with cream consumption. The cause of the outbreak was handling of the cream by an infected cook during preparation.

9

Milk Quality and Marketing

Milk is sold as sour milk, butter or ghee and some cheese. Very little milk is sold per household due to low productivity or lack of markets. The little milk sold directly to consumers and some to milk vendors who sell to consumers but such milk is very much prone to milk adulturation by water, cow's urine, coconut milk or cassava milk.

There is a lot of milk wastage especially during the rainy season when the supply is high and marketing conditions are poor.

In recent years there has emerged a strong argument advocating the use of decentralised rural based milk processing as an appropriate system of milk marketing in a situation where smallholder milk production is thinly scattered and situated far away from urban markets.

LIQUID MILK

Apart from boiling of milk before consumption there are no known traditional technologies for preservation of liquid milk through heat processing in Tanzania. Indeed is has been reported that in some pastoral communities it is a taboo to heat milk (Shallo, 1987). Table gives the technical possibilities of applying conventional milk technology processing in rural milk processing.

It is obvious that for technical and economic reasons, technologies involved in liquid milk processing (refrigeration, steam generation, pasteurization, sterilization/aseptic packing) are not immediately accessible to the small scale milk processor.

It follows then that the small scale manufacturer must undertake the processing of preserved dairy products with long shelf life at ambient temperatures such that marketing is done only once or twice a week.

Some simple equipment could be used in decentralized rural dairies in developing countries (Bachmann, 1979; Kurwijila, 1986; Schulthess, 1990). The setting up of such rural industries would create employment in the rural areas, promote local skills, as some of the equipment could be made locally, and stimulate growth of production of milk by providing an assured market for local milk.

FERMENTED MILK

Production of fermented milk is widely spread in Tanzania. In most tribes, milk is allowed to ferment naturally without addition of starter culture. Milk is accumulated over a number of days in a clay pot vessel of 2 - 4 litres volumetric capacity or a bottle gourd (O'Mahony and Bekele, 1985) and allowed to develop acidity up to 1% lactic acid.

In cold weather, the container is kept near the fire to keep it warm. Soured milk is consumed as a side dish. If not consumed, sour milk is churned into butter. The shelf life of naturally fermented or sour milk is only 72 hours at ambient temperature (Ali, 1990 and Ryoba and Kurwijila, 1992). The disadvantage of this type of milk is that since it is not boiled there is always a danger of people getting TB or Brucellosis.

Among the pastoralists like the Maasai and others, milk vessels are smoked using specific wood which produce embers in the fire (Shallo, 1987; Kurwijila, 1988b). Smoke application appears to improve the colour, flavour and extend the shelf life up to 20 days to 3 months (Bekele and Kasaye, 1987 and Ryoba and Kurwijila, 1992). Preparation of fermented milk and smoking of milk vessel have been discussed in details by Shallo and Hansen, (1987). Fidder (1966) and Knowles (1975) reports that the smoke contains primarly phenols of about 21 components as identified by the gas chromatography which have the ant bacterial effect in smoked milk thus prologing the shelf life of milk (Ryoba and Kurwijila, 1992).

Apparently no research has been done in identifying, characterizing and classifying the plants and no analysis of the smoke components and its effects on the milk components has been done.

Fidder (1966) reports that methods of generating the smoke, collecting the phenolic fractions and identifying the components have not been uniform or reproducible. The correlation of results from the investigations to obtain a meaningful information about the desirable phenol composition of smoke and the best parameters for its production is difficult. It is therefore a big challenge to the African researchers.

Natural fermentation of milk exploits lactic acid producing Streptococci and Lactobacilli which are present in any milk and which generally suppress spoilage and pathogenic organisms.

The processing steps of heating milk followed by cooling and inoculation of starter culture to achieve a controlled fermentation is not done traditionaly in Tanzania. Heating kills undesirable organisms and also concentrates the milk" giving the fermented milk a heavier body (Fox, 1967). The use of pure cultures enables different acid flavours to be developed leading to a range of fermented (cultured) products. Since cultured milk has better keeping quality than fresh milk and better product consistency than naturally soured milk, their introduction to smallholder processing systems should be considered. Standard technology of fermented milk manufacture is first to boil at 80 - 90°C for 5-30 minutes to reduce the microbial population. It is then cooled, inoculated with starter culture or previously fermented milk and incubated at 30°C to 38°C for 4-16 hours.

Higher temperatures cause wheying off while lower temperatures result in slow fermentation and may lead to the establishment of undesirable micro-organisms. Fermented milk can be made in simple containers of any size.

PROCESSING OPTIONS FOR DEVELOPMENT

Milk, as it comes from the cow, has a temperature of about 38°C. Bacteria multiply rapidly in warm milk (Robinson, 1983). In sub-Saharan Africa, where the temperature, and often humidity, are high for most of the year, raw milk spoils quickly unless it is cooled or treated with a preservative (Korhonen, 1980).

Preservatives are not readily available in rural areas and cooling systems are not feasible due to water shortages in many parts of the subcontinent. This makes collection of milk from rural areas counter-productive, as the milk quality will have deteriorated below acceptable standards by the time it reaches the centralised processing plant.

In addition, milk collection systems in sub-Saharan Africa are difficult to operate, mainly because of the seasonality of milk supply, small quantities of milk produced and infrastructural problems.

Stimulating efficient on-farm or village-level processing therefore appears to be a better strategy for dairy development in rural areas. The selection of products for small-scale manufacture will depend on the quantity of milk available for processing, the technical skills and equipment needed, marketing infrastructure and consumer preferences.

Given the small amounts of surplus milk produced by smallholders, and the scarcity of small-scale processing techniques, ILCA has tested a number of technologies which could be adopted by small-scale milk producers. Natural fermentation of milk is suitable for on-farm processing, since it is best achieved with small quantities of milk (1 to 2 litres/day) while the acidity developed preserves milk solids (Kosikowski, 1982) and improves churnability (Ephraim Bekele and O'Mahony, unpublished data).

ILCA studies have shown that as little as 0.1–0.2% fat (determined by Gerber analysis) remains in the buttermilk when sour whole milk is churned at about 14 to 16°C (O'Mahony and Ephraim Bekele, 1985a).

The equipment used to make butter from sour whole milk is available locally, and ILCA has designed a simple detachable agitator to improve churning efficiency. The butter produced can be sold fresh or salted and preserved for subsequent sale, or it can be converted into ghee. Residual liquid after butter making is highly nutritious (since it contains all the non-fat milk solids) and can be consumed without further treatment. Caseins are often recovered from the liquid in the form of ayib which is consumed fresh or preserved by salting or reducing its moisture content. Whey proteins are utilised as animal feed or in human diets. Village-level processing can be undertaken with milk volumes of up to 500 litres/ day. The recommended equipment includes hand-driven milk separators and wooden churns.

ILCA's experience to date has shown that this equipment is quickly adopted in areas where sufficient milk is available for processing. Large yields of high-quality butter are being produced at 15 producer cooperatives throughout Ethiopia, using milk separation and cream churning. Up to 200 litres of milk are being processed daily at each unit.

Centralised village processing of milk supplied by individual smallholders is presently being investigated. It is too early to comment on the effectiveness of this system in sub-Saharan Africa, but experience in India (Mogens Jul, 1977) shows that the system could be successful. Small-scale centralised milk processing reduces labour requirements on the farm, provides a nucleus for sale of farm inputs and strengthens the marketing capability of individual milk producers.

The manufacture of cheese is difficult with small quantities of milk. However, a village cooperative could make cheese and thus widen its product range.

The feasibility of making cheese by village cooperatives is now being investigated in Ethiopia. Preliminary observations indicate that cheese making may be a suitable option for rural cooperatives that are far from all-weather roads, and therefore have to accept a lower price for their butter, and which have difficulties marketing ayib locally.

The cheese varieties discussed in this paper are more stable than butter. Their manufacture would give farmers greater independence from local traders who exploit the fact that butter has to be sold before it develops rancidity above the acceptable level.

Research on the economics of milk processing by village cooperatives and the potential market for cheese in rural areas of Ethiopia is under way in order to assess the viability of small-scale cheese production.

FERMENTED MILKS

DEFINITIONS

Fermented milks have been produced by traditional methods for many centuries, and there are several hundred such products recorded around the world. They are produced as a result of microbial 'souring' of milk, usually cows' milk, but also the milk of other species, including sheep, goat, horse and buffalo. Most are very similar, both in terms of their characteristics, and in the technology used to produce them. Many fermented milk products are distinguished only by their region of origin, and very few have become commercially important.

Interest in these products, particularly yoghurt, has grown rapidly since the development of flavoured and fruit yoghurts in Europe in the late 1950s, and more recently as a result of the growing demand for, and marketing of, fermented milks as healthpromoting foods.

Fermented milks can be conveniently classified on the basis of the type of fermentation they undergo, as lactic, yeast-lactic and mould-lactic. Lactic fermentation products can be further classified, depending on the characteristics of the lactic microflora, as mesophilic, thermophilic and probiotic or therapeutic.

LACTIC ACID FERMENTATON

Lactic acid fermentation is caused by some fungi and bacteria. The most important lactic acid producing bacteria is Lactobacillus. Other bacteria which produce lactic acid include:

- leuconostoc mesenteroides
- pediococcus cerevisiae
- streptococcus lactis
- bifidobacterium bifidus.

Lactic acid fermentation is used throughout the world to produce speciality foods:

- Western world: yogurt, sourdough breads, sauerkraut, cucumber pickles and olives
- Middle East: pickled vegetables
- Korea: kimchi (fermented mixture of Chinese cabbage, radishes, red pepper, garlic and ginger)
- Russia: kefir
- Egypt: laban rayab and laban zeer (fermented milks), kishk (fermented cereal and milk mixture)
- Nigeria: gari (fermented cassava)
- South Africa : magou (fermented maize porridge)
- Thailand : nham (fermented fresh pork)
- Philippines : balao balao (fermented rice and shrimp mixture)

The presence of lactic acid, produced during the lactic acid fermentation is responsible for the sour taste and for the improved microbiological stability and safety of the food. This lactic acid fermentation is responsible for the sour taste of dairy products such as cheese, yoghurt and kefir.

Lactic acid fermentation also gives the sour taste to fermented vegetables such as traditionally cultured sauerkraut and pickles. The sugars in the cabbage are converted into lactic acid and serve as a preservative.

Yogurt fermentation

Yogurt is made by fermenting milk with friendly bacteria, mainly Lactobacillus bulgaricus and Streptococcus thermophilus. Yogurt fermentation was invented probably by accident by Balkan tribes thousands of years ago. Yogurt remained mainly a food of eastern Europe until the 1900s, when the biologist Mechnikov created the theory that lactobacillus bacteria in yogurt are responsible for the unusually long lifespans of the Bulgar people. The milk sugar or lactose is fermented by these bacteria to lactic acid which causes the characteristic curd to form.

The acid also restricts the growth of food poisoning bacteria. During the yogurt fermentation some flavours are produced, which give yogurt its characteristic flavour. Yoghurt can easily be made at home using a live yogurt as the starter culture.

To make you own yogurt use the following process. Bring the milk (or soymilk) to boiling point and cool down to 40- 45°C. Pour this milk in a sterile container and and per liter milk about 100 ml live yogurt. Mix with a sterile spoon and incubate at 40-44°C during 4 to 6 hours or until the yogurt is set. Put the yogurt in the refrigerator. If you worked under hygienic conditions, you can use your own yogurt as a starter for your next batch.

Magou fermentation

Magou is very popular in South Africa, especially among the Bantu people. Magou is a lactic acid fermented porridge made from maize. To make magua a 10 percent maize meal slurry is cooked, cooled and inoculated with wheat flour, which contains the bacteria. Magou is also produced on industrial scale and is then packed in cartons. In the industrial process the magou is inoculated with lactobacillus delbreuckii cultures.

Kefir fermentation

Kefir fermentation is similar to yogurt fermentation. Yogurt is only fermented by bacteria but kefir fermentation involves the help of bacteria as well as yeasts.

These yeast produce some alcohol and carbon dioxide, which gives kefir its typical fizzy aspect. Kefir is inoculated with special kefir grains. These grains are mixtures of bacteria and yeasts in a matrix of proteins, lipids and carbohydrates. Kefir fermentation is done at room temperature, which makes the process easier. On the other hand, not everyone likes the taste of kefir.

MESOPHILIC

The genera of microorganisms that fall into this category include *Lactococcus*, *Leuconostoc* and *Pediococcus*. The optimal growth temperature is between 25-30 °C.

Traditional or Natural Buttermilk

Traditional or natural buttermilk is made from the liquid produced during butter production using a starter culture mixture of *Lactococcus* spp. and *Leuconostoc mesenteroides* subsp. *cremoris*.

Cultured Buttermilk

Cultured buttermilk is also produced mostly using a mixed culture of *L. lactis* subsp. *lactis*, *L. lactis* subsp. *cremoris* and the flavour-producing organisms *L. lactis* biovar *diacetylactis* and *L. mesenteroides* subsp. *cremoris*. It is traditionally made from skimmed milk.

Ymer is similar to cultured buttermilk, but differs in the sequence of the manufacturing stages.

Nordic Sour Milks

Nordic sour milks such as Filmjolk and Nordic ropy milk are made using slimeproducing *Bacterium lacticus longi*, a synonym of *Lactococcus* spp. The slimy or ropy consistency of the products is also attributed to Butterwort leaves, which are rubbed on the interior of the pails.

Cultured Cream

Cultured cream or sour cream is made using the same starter cultures as cultured buttermilk, but has a much higher fat content.

Miscellaneous Products

Miscellaneous products include a range of traditional products that depend on spontaneous fermentation by naturally present lactic acid bacteria in milk. Maziwa lala is made using the same starter culture mixture as buttermilk, but is then sweetened.

Susa, made from camel's milk is fermented using hetero-fermentative mesophilic starter cultures.

Lben is similar to buttermilk but its production involves spontaneous fermentation.

The microflora of this product mainly consists of *L. lactis* biovar *diacetylactis*, *Leuconostoc lactis*, *L. mesenteroides* subsp. *cremoris* and *Leuconostoc mesenteroides* subsp. *dextranicum*; lactobacilli, yeast, mould and coliforms are also present.

THERMOPHILIC

This category encompasses those starter cultures whose growth optimum is between 37 and 45°C.

The genera of microorganisms that fall into this category include *Streptococcus* and *Lactobacillus*.

Yoghurt

Yoghurt is a term used to describe a wide range of related products, which may be classified according to legal standards, gel type and whether or not they are flavoured or if they are subjected to a further process. The usual starter culture employed to produce yoghurt is a mixture of *Streptococcus thermophilus* and *Lactobacillus delbrueckii* subsp. *bulgaricus*.

Acid Buttermilk

Acid buttermilk, also known as Bulgarian buttermilk is made using *L. delbrueckii* subsp. *bulgaricus* as the starter culture. *Str. thermophilus* or a cream culture may also be included in the starter culture.

PROBIOTIC OR THERAPEUTIC

LAB such as enterococci, lactococci, propionibacteria, *Leuconostoc*, and pediococci are used as probiotics, but the principal organisms are of the bacterial genera *Lactobacillus* and *Bifidobacterium*.

Yakult

Yakult is a term for a group of therapeutic products originating from Japan. The starter culture used is *Lactobacillus casei* subsp. *casei*, an organism naturally present in the normal intestinal microflora of humans.

The organism is a probiotic strain that is thought to have a beneficial effect on the host, by improving the intestinal microbial balance.

The positive health benefits of probiotics are reported to be of particular value in the treatment of diseases that result in a disturbance of the intestinal microflora.

Acidophilus Milk

Acidophilus milk is a traditional therapeutic milk product popular in eastern Europe, but now attracting more attention elsewhere for its perceived beneficial properties.

It may be made from skimmed or whole milk, and the starter organism is *Lactobacillus acidophilus*.

'Bio' Yoghurts

'Bio' yoghurts are made by very much the same process as traditional yoghurt, and are very similar products, but usually use a mixed starter culture consisting of probiotic strains. *Bifidobacterium* spp. are often used, especially *Bifidobacterium bifidum*, and *Bifidobacterium longum*, together with lactobacilli, such as *L. casei* and *L. acidophilus*.

These organisms are all found in the normal intestinal microflora and are considered to have a beneficial effect on human health.

YEAST-LACTIC FERMENTATIONS

Mesophilic LAB, thermophilic LAB and yeast are the main fermentation genera.

KEFIR

Kefir is a rather foamy and effervescent fermented milk that contains about 1% lactic acid and 0.5-1.0% alcohol, and is popular in eastern Europe and Mongolia. The starter culture consists of small, white 'kefir grains', about 2-10 mm in diameter.

These grains contain a complex and quite variable microbial community, but little is known about how they develop. The grains usually contain LAB such as *Lactobacillus* spp. plus *Lactococcus* spp., *Leuconostoc* spp., and *Str. thermophilus*, acetic acid bacteria, contaminants such as mould, and a number of yeast species such as *Saccharomyces and Kluyveromyces*, but the principal yeast species present is *Candida kefir*.

KOUMISS

Koumiss is traditionally made in central Asia from mares' milk, but is now often made from skimmed, or whole cows' milk with added sugar. Starter cultures contain LAB such as lactobacilli, strains of lactose-fermenting yeasts, non-lactose-fermenting and non-carbohydrate-fermenting yeasts. The finished product contains lactic acid, alcohol and carbon dioxide, producing a slightly effervescent drink.

MISCELLANEOUS PRODUCTS

Miscellaneous products such as acidophilus-yeast milk fall under the yeast-lactic group of fermented products, but little is known about the technology of these beverages.

MOULD-LACTIC FERMENTATIONS

Mesophilic LAB and mould are the genera responsible for fermentation.

VILLI

Villi is a fermented milk product from Finland, which is made from whole milk, using a starter culture of *L. lactis* subsp. lactis biovar *diacetylactis*, *L. mesenteroides* subsp. *cremoris*, and the mould *Geotrichum candidum*. The mould grows on the layer of fat that forms on the top of the product and produces a felt of mycelium.

INITIAL MICROFLORA

The initial microflora of fermented milk products is determined largely by the microflora of the whole and skimmed milks from which they are made.

PROCESSING AND ITS EFFECTS ON THE MICROFLORA

Although there is a very wide range of fermented milk products, the manufacturing technology used is generally very similar. The principal differences are in the starter cultures used, the composition and treatment of the milk, and the fermentation conditions.

Therefore, for the purposes of this chapter, yoghurt manufacture is used as a representative example of fermented milk processes, since yoghurt is the most commercially important of these products. An outline of the process is depicted in Figure.

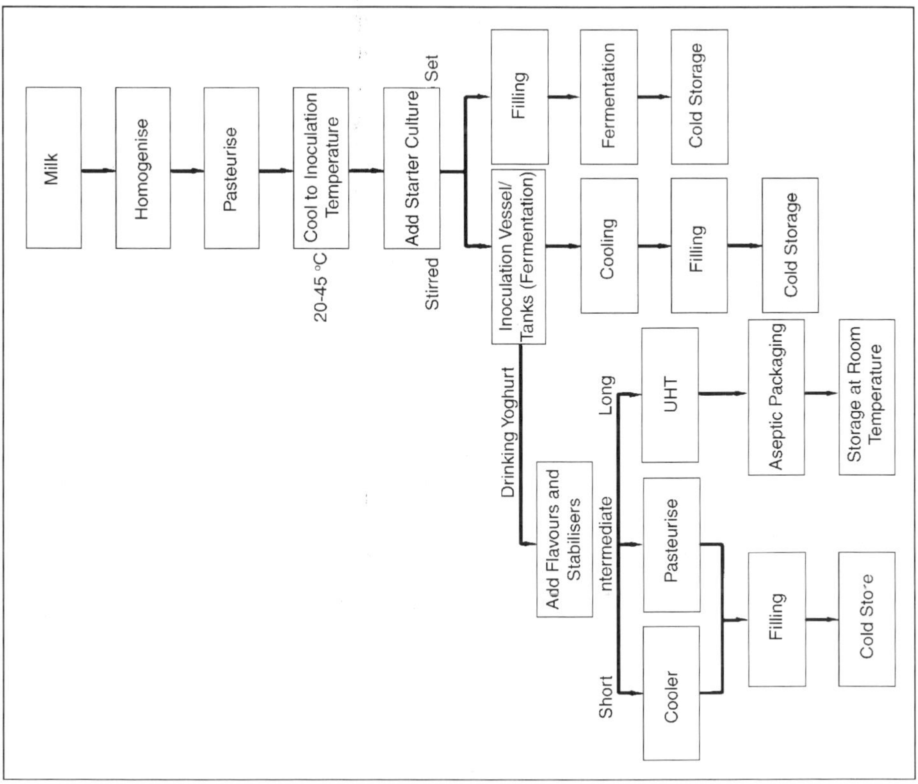

Fig. Production of Stirred and Set Fermented Milk

INITIAL PROCESSING

Several different varieties of yoghurt are produced, but the three main types are: set; stirred; and drinking yoghurt. Yoghurt is most commonly made from cows' milk, but may also be produced from the milk of sheep, goats, and, occasionally, other animal species. The composition of yoghurt varies slightly, and in some countries is regulated by legislation. Both whole milk and reduced-fat milks are used to produce yoghurt, but reduced-fat products have the largest market share in most countries. A fat content of approximately 1.5% is typical for a low-fat yoghurt, and the milk is usually standardised to control the final fat content.

The protein composition and quality of the milk are important, since they may have a significant effect on texture.

Only milk of good microbiological quality should be used, in order to avoid problems of proteolysis associated with bacterial activity, and the production of bacterial proteases by psychrotrophs.

These enzymes may significantly alter the physical properties of the yoghurt, and cause defects.

The milk-solids-not-fat content of the milk is usually increased to give a higher viscosity in the finished product. This may be done by fortification with non-fat dried milk or other dairy powders, or by concentration methods, such as evaporation under vacuum, or by membrane filtration. For most yoghurt, an MSNF level of about 15% is typical, but for drinking yoghurt, levels of less than 11% are preferred.

The milk is usually then filtered, to remove undissolved particles, de-aerated to provide conditions that favour rapid starter growth, and homogenised to improve texture and help prevent syneresis. Stabilisers may also be added to stirred yoghurts to improve viscosity and further reduce the likelihood of syneresis. Pre-gelatinised starch or plant gums are the most commonly used stabilisers.

HEAT TREATMENT

A heat treatment is generally applied to milk for yoghurt manufacture. A process of 80-85°C for 30 minutes is typical for batch processes, but, for continuous processes, a heat treatment of 90-95°C for 5-10 minutes is more usual.

In some cases, a full ultra high temperature process may be applied. This relatively severe heat treatment has a number of effects.

Vegetative bacterial cells, which may include pathogens such as *Salmonella*, are killed, leaving only heat-resistant bacterial spores. Non-pathogenic organisms that might interfere with the growth of the starter culture are therefore reduced to very low levels. The heat treatment also has a significant effect on the final viscosity of the yoghurt, and improves texture by causing denaturation of the whey proteins and the three-dimensional aggregation of casein molecules.

The oxygen concentration of the milk is also further reduced, improving conditions for starter culture growth, since the starter organisms are generally microaerophilic.

Finally, starter culture activity can either be stimulated or inhibited due to the breakdown products of heat-damaged milk proteins.

FERMENTATION

After heat treatment, the milk is cooled to 30 or 45°C and is then inoculated with the starter culture. Most commercial yoghurt production now uses a mixed inoculum containing defined strains of *Str. thermophilus* and *L. delbrueckii* subsp. *bulgaricus*.

During fermentation, the two starter organisms grow synergistically. Initially, *Str. thermophilus* grows rapidly and produces lactic acid. As the acidity rises, so *L. delbreuckii* subsp. *bulgaricus* becomes more active and produces more acid along with aroma compounds.

It has been found that the growth of *Str. thermophilus* stimulates the growth of the *Lactobacillus*, probably by the production of a growth factor, thought to be formic acid, which stimulates proteolytic activity.

Lactobacillus growth is further stimulated by the production of low levels of carbon dioxide from urea. The lactobacilli, in turn, stimulate the growth of *Str. thermophilus* by releasing peptides and, to lesser extent, amino acids as by-products of proteolysis.

The specific strains used are chosen for their effect on product flavour and texture, and strains may be used in rotation to prevent occasional problems with bacteriophage infection.

Good hygienic practices after pasteurisation are required to prevent build-up of bacteriophage pools in stagnant whey. It is important that the inoculum contains a balanced population of the two organisms, usually a 1:1 ratio. Both liquid and freeze-dried cultures are used. Most commercial manufacturers use an inoculum of about 2-3% v/v.

Incubation of the inoculated milk at 40-45°C produces a rapid fermentation, which is normally complete within 3-4 hours. During this time, lactic acid is produced, giving an eventual acidity of 0.9-0.95% and a pH of approximately 4.6-5.0.

The starter organisms may also produce aroma compounds, such as acetaldehyde, acetone, acetoin and diacetyl, and exopolysaccharides, which improve texture and viscosity, although if too much extracellular material is produced, a 'ropy' texture fault may result. Set yoghurt is fermented in the final container, but stirred yoghurt is fermented in bulk, and stirred slowly, for only a few minutes, during the process.

COOLING AND PACKING

When fermentation is complete, the yoghurt is initially cooled to about 15-20°C, to minimise further acid production. At this point, sweeteners, flavours and/or fruit purees may be added. These additions must be of good microbiological quality, since, for most yoghurts, no further processing is applied. The product is then dispensed into the final containers and further cooled to < 5°C.

A shelf life of about 3 weeks at this temperature is typical, although acid continues to be produced during this time and may affect flavour. Some yoghurt is heat treated after fermentation to destroy starter organisms, and this increases shelf life to several months. Greek-style, concentrated yoghurts are produced by further separation of the yoghurt after fermentation to increase the fat and solids content. Yoghurt may also be frozen, or dried for use as an ingredient.

Bibliography

Alan H. Varnam and Jane P. Sutherland: *Milk and Milk Products : Technology, Chemistry and Microbiology*, Springer Publication, London, 2010.

Andrew L. Winton and Kate Barber Winton: *Milk and Milk Products*, Agrobios Publication, Jaipur, 2002.

Aysha Aziz Faridi: *Dairy Microbiology*, Random Publications, Delhi, 2012.

B. K. Mishra: *Dairy and Food Processing Industry : Recent Trends (2 Vols-Set)*, Biotech Books, New Delhi, 2014.

Dilip Shah: *Co-Operativization Liberalization and Dairy Industry in India*, ABD Publishers, Delhi, 2000.

Dinesh Arora: *Biotech's Dictionary of Dairy Science*, Biotech Books, New Delhi, 2003.

G. Sutherland Thomson: *Grading Dairy Produce*, Medi World Press, Kerala, 1995.

Gaurav Bhardwaj: *Applied Dairy Microbiology*, Sonali Publications, Delhi, 2012.

H.A. Modi: *Dairy Microbiology*, Aavishkar Publishers, Delhi, 2009.

H.A. Modi: *Fermented Milk Products*, Aavishkar Publication, Delhi, 2011.

Harmeet Singh: *Dairy Farming*, APH Publication, Delhi, 2011.

J David: *Contemporary Trends in Dairy Plant Management*, Kitab Mahal Agencies, Delhi, 2007.

K. Singh: *Dairy Microbiology*, Oxford Book Company, Delhi, 2012.

M K Rao: *Food and Dairy Microbiology*, Manglam Publication, Delhi, 2007.

M. Singh, D.K. Sharma and U.K. Mishra: *Organic Dairy Farming*, Satish Serial Publication, Delhi, 2011.

Mangala Kango: *Milk and Milk Products*, RBSA Publication, Jaipur, 2006.

Manoj Negi: *Career in Dairy Farming*, Abhishek Publication, Delhi, 2010.

Mohammad Raziuddin and Ashok Hembade: *Milk and Milk Products Technology*, Jaya Publishing House, Delhi, 2013.

Mohd. Kalimuddin: *Milk Analysis and Dairy Farming*, Enkay Publication, Delhi, 2012.

Mohd. Kalimuddin: *Milk and Milk Products*, Enkay Publication, Delhi, 2012.

N.S. Rathore, A.S. Jhala, G.K. Mathur and S.S. Chasta: *Consumer Awareness for Food and Dairy Products*, Apex Publication, Delhi, 2006.

P R Gupta: *Dairy India 2007: Serving the Dairy Industry Since 1983*, Dairy India Yearbook, 2007.

P. Venkateshwara Rao: *Principles of Dairy Management*, Biotcch Books, New Delhi, 2008.

Pradeep Parihar and Leena Parihar: *Dairy Microbiology*, Agrobios Publication, Jaipur, 2008.

Purshotam Kaushik: *Indigenous Medicinal Plants Including Microbes and Fungi*, Today and Tomorrows Publication, Delhi, 2009.

Rajiv K Sinha and Rohit Sinha: *Environmental Biotechnology : Role* of Plants, Microbes and Earthworms in Environmental

Management and Sustainable Development, Aavishkar Publication, Delhi, 2008.

Rameshwar Singh, S.K. Tomar and Gunjan Goel: *Applied Dairy and* 294 Dairy Microbiology

Food Microbiology : Contemporary Reviews, Agrotech Publication, Delhi, 2005.

Robert Jennings: *Organic Dairying and Integrated Agriculture*, Dominant Publishers, Delhi, 2012.

S N Jha: *Dairy and Food Processing Plant Maintenance : Theory and Practice*, International Book Distributing, Delhi, 2006.

S. Sundara Rajan: *Genetics and Physiology of Microbes*, Anmol Publication, 2003.

S.K. Soni: *Microbes : A Source of Energy for 21 Century*, New India Publication, Delhi, 2007.

S.K. Tomar, R. Singh, A.K. Singh, Sumit Arora and R.R.B Singh: *Functional Dairy Foods Concepts and Applications*, Satish Serial Publication, Delhi, 2011.

S.N. Mahindru: *Milk and Milk Products*, APH Publication, Delhi, 2012.

Shyam Kishore Singh: *Dairy Farming*, Alfa Publications, Delhi, 2012.

Sudhir Chandra, K.K. Khanna and H.K. Kehri: *Microbes and Man*, Bishen Singh Mahendra Pal Singh Publishers, Dehra Dun, 1995.

Uma Shankar Singh: *Dairy Farming*, Anmol Publication, Delhi, 2008.

Varnam: *Milk and Milk Products: Technology, Chemistry, and Microbiology*, Springer Publication, London, 2003.

Index